Introduction to Modeling and Simulation of Technical and Physical Systems with Modelica

Introduction to Modeling and Simulation of Technical and Physical Systems with Modelica

Peter Fritzson

IEEE Press

A John Wiley & Sons, Inc., Publication

For further information visit: the book web page http://www.openmodelica.org, the Modelica Association web page http://www.modelica.org, the authors research page http://www.ida.liu.se/labs/pelab/modelica, or home page http://www.ida.liu.se/~petfr/, or email the author at peter.fritzson@liu.se. Certain material from the Modelica Tutorial and the Modelica Language Specification available at http://www.modelica.org has been reproduced in this book with permission from the Modelica Association under the Modelica License 2 Copyright © 1998–2011, Modelica Association, see the license conditions (including the disclaimer of warranty) at http://www.modelica.org/modelica-legal-documents/ModelicaLicense2.html. Licensed by Modelica Association under the Modelica License 2.

Modelica® is a registered trademark of the Modelica Association. MathModelica® is a registered trademark of MathCore Engineering AB. Dymola® is a registered trademark of Dassault Systèmes. MATLAB® and Simulink® are registered trademarks of MathWorks Inc. Java™ is a trademark of Sun MicroSystems AB. Mathematica® is a registered trademark of Wolfram Research Inc.

For general information on our other products and services or for technical support, please contact our Customer Care Department within the United States at (800) 762-2974, outside the United States at (317) 572-3993 or fax (317) 572-4002.

Wiley also publishes its books in a variety of electronic formats. Some content that appears in print may not be available in electronic formats. For more information about Wiley products, visit our web site at www.wiley.com.

Library of Congress Cataloging-in-Publication Data:

Fritzson, Peter A., 1952-
  Introduction to modeling and simulation of technical and physical systems with Modelica /
Peter Fritzson.
    p. cm.
  Includes bibliographical references and index.
  ISBN 978-1-118-01068-6 (cloth)
1. Systems engineering—Data processing. 2. Computer simulation. 3. Modelica. I. Title.
  TA168.F76 2011
  003'.3—dc22

                                                                2011002187

Contents

Preface

This book teaches the basic concepts of modeling and simulation and gives an introduction to the Modelica language to people who are familiar with basic programming concepts. It gives a basic introduction to the concepts of modeling and simulation, as well as the basics of object-oriented component-based modeling for the novice. The book has the following goals to be:

- A useful textbook in introductory courses on modeling and simulation.
- Easily accessible for people who do not previously have a background in modeling, simulation and object orientation.
- A basic introduction of the concepts of physical modeling, object-oriented modeling, and component-based modeling.
- A demonstration of modeling examples from a few selected application areas.

The book contains examples of models in a few different application domains, as well as examples combining several domains.

All examples and exercises in this book are available in an electronic self-teaching material called DrModelica, based on this book and the more extensive book *Principles of Object-Oriented Modeling of Simulation with Modelica 2.1* Fritzson (2004), for which an updated version is planned. DrModelica gradually guides the reader from simple introductory examples and exercises to more advanced ones. Part of this teaching material can be freely downloaded from the book's website, www.openmodelica.org, where additional (teaching) material related to this book can be found.

ACKNOWLEDGEMENTS

The members of the Modelica Association created the Modelica language and contributed many examples of Modelica code in the Modelica Language Rationale and Modelica Language Specification (see http://www.modelica.org), some of which are used in this book. The members who contributed to various versions of Modelica are mentioned further below.

First, thanks to my wife, Anita, who has supported and endured me during this writing effort.

Special thanks to Peter Bunus for help with model examples, some figures, MicroSoft Word formatting, and for many inspiring discussions. Many thanks to Adrian Pop, Peter Aronsson, Martin Sjölund, Per Östlund, Adeel Asghar, Mohsen Torabzadeh-Tari, and many other people contributing to the OpenModelica effort for a lot of work on the OpenModelica compiler and system, and also to Adrian for making the OMNotebook tool finally work. Many thanks to Hilding Elmqvist for sharing the vision about a declarative modeling language, for starting off the Modelica design effort by inviting people to form a design group, for serving as the first chairman of Modelica Association, and for enthusiasm and many design contributions including pushing for a unified class concept. Also thanks for inspiration regarding presentation material including finding historical examples of equations.

Many thanks to Martin Otter for serving as the second chairman of the Modelica Association, for enthusiasm and energy, design and Modelica library contributions, for two of the tables and some text in Chapter 5 on Modelica libraries from the Modelica Language Specification, as well as inspiration regarding presentation material. Thanks to Jakob Mauss who made the first version of the Glossary, and to Members of Modelica Association for further improvements.

Many thanks to Eva-Lena Lengquist Sandelin and Susanna Monemar for help with the exercises, and for preparing the first version of the DrModelica interactive notebook teaching material which makes the examples in this book more accessible for interactive learning and experimentation.

Thanks to Peter Aronsson, Jan Brugård, Hilding Elmqvist, Vadim Engelson, Dag Fritzson, Torkel Glad, Pavel Grozman, Emma Larsdotter Nilsson, Håkan Lundvall, and Sven-Erik Mattsson for constructive comments, on parts of the book. Thanks to Hans Olsson and Martin Otter who edited recent versions of the Modelica Specification. Thanks to all members of PELAB and employees of MathCore Engineering, who have given comments and feedback.

Linköping, Sweden　　　　　　　　　　　　　　PETER FRITZSON
May 2011

CHAPTER 1

Basic Concepts

It is often said that computers are revolutionizing science and engineering. By using computers we are able to construct complex engineering designs such as space shuttles. We are able to compute the properties of the universe as it was fractions of a second after the big bang. Our ambitions are ever-increasing. We want to create even more complex designs such as better spaceships, cars, medicines, computerized cellular phone systems, and the like. We want to understand deeper aspects of nature. These are just a few examples of computer-supported modeling and simulation. More powerful tools and concepts are needed to help us handle this increasing complexity, which is precisely what this book is about.

This text presents an object-oriented component-based approach to computer-supported mathematical modeling and simulation through the powerful Modelica language and its associated technology. Modelica can be viewed as an almost universal approach to high-level computational modeling and simulation, by being able to represent a range of application areas and providing general notation as well as powerful abstractions and efficient implementations. The introductory part of this book, consisting of the first two chapters, gives a quick overview of the two main topics of this text:

- Modeling and simulation
- The Modelica language

Introduction to Modeling and Simulation of Technical and Physical Systems with Modelica,
First Edition. By Peter Fritzson
© 2011 the Institute of Electrical and Electronics Engineers, Inc. Published 2011 by John Wiley & Sons, Inc.

The two subjects are presented together since they belong together. Throughout the text Modelica is used as a vehicle for explaining different aspects of modeling and simulation. Conversely, a number of concepts in the Modelica language are presented by modeling and simulation examples. The present chapter introduces basic concepts such as *system, model*, and *simulation*. Chapter 2 gives a quick tour of the Modelica language as well as a number of examples, interspersed with presentations of topics such as object-oriented mathematical modeling. Chapter 3 gives an introduction to the Modelica class concept, whereas Chapter 4 introduces modeling methodology for continuous, discrete, and hybrid systems. Chapter 5 gives a short overview of the Modelica Standard Library and some currently available Modelica model libraries for a range of application domains. Finally, in two of the appendices, examples are presented of textual modeling using the OpenModelica electronic book OMNotebook tool, as well as very simple graphical modeling.

1.1 SYSTEMS AND EXPERIMENTS

What is a system? We have already mentioned some systems such as the universe, a space shuttle, and the like. A system can be almost anything. A system can contain subsystems that are themselves systems. A possible definition of system might be:

- A system is an object or collection of objects whose properties we want to study.

Our wish to study selected properties of objects is central in this definition. The "study" aspect is fine despite the fact that it is subjective. The selection and definition of what constitutes a system is somewhat arbitrary and must be guided by what the system is to be used for.

What reasons can there be to study a system? There are many answers to this question but we can discern two major motivations:

- Study a system to understand it in order to build it. This is the engineering point of view.
- Satisfy human curiosity, for example, to understand more about nature—the natural science viewpoint.

1.1.1 Natural and Artificial Systems

A system according to our previous definition can occur naturally, for example, the universe, it can be artificial such as a space shuttle, or a mix of both. For example, the house in Figure 1.1 with solar-heated warm tap water is an artificial system, that is, manufactured by humans. If we also include the sun and clouds in the system, it becomes a combination of natural and artificial components.

Even if a system occurs naturally, its definition is always highly selective. This is made very apparent in the following quote from Ross Ashby (1956, p. 39):

> *At this point, we must be clear about how a* system *is to be defined. Our first impulse is to point at the* pendulum *and to say "the system is that thing there." This method, however, has a fundamental disadvantage: every material object contains no less than an infinity of variables, and therefore, of possible systems. The real pendulum, for instance, has not only length and position; it has also mass, temperature, electric conductivity, crystalline structure, chemical impurities, some radioactivity, velocity, reflecting power, tensile strength, a surface film of moisture, bacterial contamination, an optical absorption, elasticity, shape, specific gravity, and so on and on. Any suggestion that we should study all the facts is unrealistic, and actually the attempt is never made.*

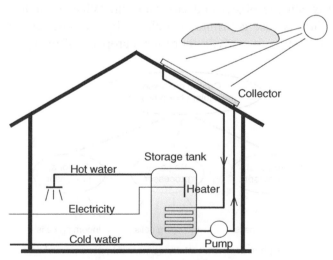

Figure 1.1   A system: a house with solar-heated warm tap water, together with clouds and sunshine.

What is necessary is that we should pick out and study the facts that are relevant to some main interest that is already given.

Even if the system is completely artificial, such as the cellular phone system depicted in Figure 1.2, we must be highly selective in its definition, depending on what aspects we want to study for the moment.

An important property of systems is that they should be *observable*. Some systems, but not large natural systems like the universe, are also *controllable* in the sense that we can influence their behavior through inputs, that is:

- The *inputs* of a system are variables of the environment that influence the behavior of the system. These inputs may or may not be controllable by us.
- The *outputs* of a system are variables that are determined by the system and may influence the surrounding environment.

In many systems the same variables act as *both inputs and outputs*. We talk about *acausal* behavior if the relationships or influences between variables do not have a causal direction, which is the case for relationships described by equations. For example, in a mechanical system the forces from the environment influence the displacement of an object, but on the other hand the displacement of the object influences the forces between the object and environment. What is input and what is output in this case is primarily a choice by the observer, guided by what is interesting to study, rather than a property of the system itself.

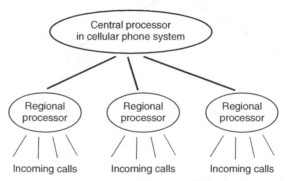

Figure 1.2   Cellular phone system containing a central processor and regional processors to handle incoming calls.

1.1.2 Experiments

Observability is essential in order to study a system according to our definition of system. We must at least be able to observe some outputs of a system. We can learn even more if it is possible to exercise a system by controlling its inputs. This process is called *experimentation*, that is:

- An *experiment* is the process of extracting information from a system by exercising its inputs.

To perform an experiment on a system, it must be both controllable and observable. We apply a set of external conditions to the accessible inputs and observe the reaction of the system by measuring the accessible outputs.

One of the disadvantages of the experimental method is that for a large number of systems many inputs are not accessible and controllable. These systems are under the influence of inaccessible inputs, sometimes called *disturbance inputs*. Likewise, it is often the case that many really useful possible outputs are not accessible for measurements; these are sometimes called *internal states* of the system. There are also a number of practical problems associated with performing an experiment, for example:

- The experiment might be too *expensive*: Investigating ship durability by building ships and letting them collide is a very expensive method of gaining information.
- The experiment might be too *dangerous*: Training nuclear plant operators in handling dangerous situations by letting the nuclear reactor enter hazardous states is not advisable.
- The *system* needed for the experiment might *not yet exist*. This is typical of systems to be designed or manufactured.

The shortcomings of the experimental method led us to the model concept. If we make a model of a system, this model can be investigated and may answer many questions regarding the real system if the model is realistic enough.

1.2 THE MODEL CONCEPT

Given the previous definitions of system and experiment, we can now attempt to define the notion of model:

- A *model* of a system is anything an "experiment" can be applied to in order to answer questions about that *system*.

This implies that a model can be used to answer questions about a system *without* doing experiments on the *real* system. Instead we perform simplified "experiments" on the model, which in turn can be regarded as a kind of simplified system that reflects properties of the real system. In the simplest case a model can just be a piece of information that is used to answer questions about the system.

Given this definition, any model also qualifies as a system. Models, just like systems, are hierarchical in nature. We can cut out a piece of a model, which becomes a new model that is valid for a subset of the experiments for which the original model is valid. A model is always related to the system it models and the experiments to which it can be subjected. A statement such as "a model of a system is invalid" is meaningless without mentioning the associated system and the experiment. A model of a system might be valid for one experiment on the model and invalid for another. The term model *validation*, see Section 1.5.3, always refers to an experiment or a class of experiment to be performed.

We talk about different kinds of models depending on how the model is represented:

- *Mental* model—a statement like "a person is reliable" helps us answer questions about that person's behavior in various situations.
- *Verbal* model—this kind of model is expressed in words. For example, the sentence "More accidents will occur if the speed limit is increased" is an example of a verbal model. Expert systems is a technology for formalizing verbal models.
- *Physical* model—this is a physical object that mimics some properties of a real system, to help us answer questions about that system. For example, during design of artifacts such as

buildings, airplanes, and so forth, it is common to construct small physical models with the same shape and appearance as the real objects to be studied, for example, with respect to their aerodynamic properties and aesthetics.

- *Mathematical* model—a description of a system where the relationships between variables of the system are expressed in mathematical form. Variables can be measurable quantities such as size, length, weight, temperature, unemployment level, information flow, bit rate, and so forth. Most laws of nature are mathematical models in this sense. For example, Ohm's law describes the relationship between current and voltage for a resistor; Newton's laws describe relationships between velocity, acceleration, mass, force, and the like.

The kinds of models that we primarily deal with in this book are mathematical models represented in various ways, for example, as equations, functions, computer programs, and the like. Artifacts represented by mathematical models in a computer are often called *virtual prototypes*. The process of constructing and investigating such models is virtual prototyping. Sometimes the term *physical modeling* is used also for the process of building mathematical models of physical systems in the computer if the structuring and synthesis process is the same as when building real physical models.

1.3 SIMULATION

In the previous section we mentioned the possibility of performing "experiments" on models instead of on the real systems corresponding to the models. This is actually one of the main uses of models, and is denoted by the term *simulation*, from the Latin *simulare*, which means to pretend. We define a simulation as follows:

- A *simulation* is an experiment performed on a model.

Analogous to our previous definition of *model*, this definition of simulation does not require the model to be represented in mathematical or computer program form. However, in the rest of this text we will concentrate on *mathematical models*, primarily those that have a computer-representable form. The following are a few examples of such experiments or simulations:

- A simulation of an industrial process such as steel or pulp manufacturing, to learn about the behavior under different operating conditions in order to improve the process.
- A simulation of vehicle behavior, for example, of a car or an airplane, for the purpose of providing realistic operator training.
- A simulation of a simplified model of a packet-switched computer network, to learn about its behavior under different loads in order to improve performance.

It is important to realize that the *experiment description* and *model description* parts of a simulation are conceptually separate entities. On the other hand, these two aspects of a simulation belong together even if they are separate. For example, a model is valid only for a certain class of experiments. It can be useful to define an *experimental frame* associated with the model, which defines the conditions that need to be fulfilled by valid experiments.

If the mathematical model is represented in executable form in a computer, simulations can be performed by *numerical experiments*, or in nonnumeric cases by *computed experiments*. This is a simple and safe way of performing experiments, with the added advantage that essentially all variables of the model are observable and controllable. However, the value of the simulation results is completely dependent on how well the model represents the real system regarding the questions to be answered by the simulation.

Except for experimentation, simulation is the only technique that is generally applicable for analysis of the behavior of arbitrary systems. Analytical techniques are better than simulation, but usually apply only under a set of simplifying assumptions, which often cannot be justified. On the other hand, it is not uncommon to combine analytical techniques with simulations, that is, simulation is used not alone but in an interplay with analytical or semianalytical techniques.

1.3.1 Reasons for Simulation

There are a number of good reasons to perform simulations instead of performing experiments on real systems:

- Experiments are too *expensive*, too *dangerous*, or the system to be investigated does *not yet exist*. These are the main difficulties of experimentation with real systems, previously mentioned in Section 1.1.2.

- The *time scale* of the dynamics of the system is not compatible with that of the experimenter. For example, it takes millions of years to observe small changes in the development of the universe, whereas similar changes can be quickly observed in a computer simulation of the universe.

- Variables may be *inaccessible*. In a simulation all variables can be studied and controlled, even those that are inaccessible in the real system.

- Easy *manipulation* of models. Using simulation, it is easy to manipulate the parameters of a system model, even outside the feasible range of a particular physical system. For example, the mass of a body in a computer-based simulation model can be increased from 40 to 500 kg at a keystroke, whereas this change might be hard to realize in the physical system.

- Suppression of *disturbances*. In a simulation of a model it is possible to suppress disturbances that might be unavoidable in measurements of the real system. This can allow us to isolate particular effects and thereby gain a better understanding of those effects.

- Suppression of *second-order effects*. Often, simulations are performed since they allow suppression of second-order effects such as small nonlinearities or other details of certain system components, which can help us to better understand the primary effects.

1.3.2 Dangers of Simulation

The ease of use of simulation is also its most serious drawback: It is quite easy for the user to forget the limitations and conditions under which a simulation is valid and therefore draw the wrong conclusions from the simulation. To reduce these dangers, one should always try to compare at least some results of simulating a model against experimental results from the real system. It also helps to be aware of the following three common sources of problems when using simulation:

- Falling in love with a model—the Pygmalion[1] effect. It is easy to become too enthusiastic about a model and forget all about the experimental frame, that is, that the model is not the real world but only represents the real system under certain conditions. One example is the introduction of foxes on the Australian continent to solve the rabbit problem, on the model assumption that foxes hunt rabbits, which is true in many other parts of the world. Unfortunately, the foxes found the indigenous fauna much easier to hunt and largely ignored the rabbits.
- Forcing reality into the constraints of a model—the Procrustes[2] effect. One example is the shaping of our societies after currently fashionable economic theories having a simplified view of reality, and ignoring many other important aspects of human behavior, society, and nature.
- Forgetting the model's level of accuracy. All models have simplifying assumptions, and we have to be aware of those in order to correctly interpret the results.

For these reasons, while analytical techniques are generally more restrictive since they have a much smaller domain of applicability, such techniques are more powerful when they apply. A simulation result is valid only for a particular set of input data. Many simulations are needed to gain an approximate understanding of a system. Therefore, if analytical techniques are applicable, they should be used instead of a simulation or as a complement.

1.4 BUILDING MODELS

Given the usefulness of simulation in order to study the behavior of systems, how do we go about building models of those systems? This

---

[1] Pygmalion is the mythical king of Cyprus who also was a sculptor. The king fell in love with one of his works, a sculpture of a young woman, and asked the gods to make her alive.

[2] Procrustes is a robber known from Greek mythology. He is known for the bed where he tortured travelers who fell into his hands: If the victim was too short, he stretched arms and legs until the person fit the length of the bed; if the victim was too tall, he cut off the head and part of the legs.

is the subject of most of this book and of the Modelica language, which has been created to simplify model construction as well as reuse of existing models.

There are in principle two main sources of general system-related knowledge needed for building mathematical models of systems:

- The collected *general experience* in relevant domains of science and technology, found in the literature and available from experts in these areas. This includes the *laws of nature*, for example, including Newton's laws for mechanical systems, Kirchhoff's laws for electrical systems, approximate relationships for nontechnical systems based on economic or sociological theories, and so on.
- The *system* itself, that is, observations of and experiments on the system we want to model.

In addition to the above system knowledge, there is also specialized knowledge about mechanisms for handling and using facts in model construction for specific applications and domains, as well as generic mechanisms for handling facts and models, that is:

- *Application expertise* —mastering the application area and techniques for using all facts relative to a specific modeling application.
- *Software and knowledge engineering* —generic knowledge about defining, handling, using, and representing models and software, for example, object orientation, component system techniques, expert system technology, and so on.

What is then an appropriate analysis and synthesis *process* to be used in applying these information sources for constructing system models? Generally, we first try to identify the main components of a system and the kinds of interaction between these components. Each component is broken down into subcomponents until each part fits the description of an existing model from some model library, or we can use appropriate laws of nature or other relationships to describe the behavior of that component. Then we state the component interfaces and make a mathematical formulation of the interactions between the components of the model.

Certain components might have unknown or partially known model parameters and coefficients. These can often be found by fitting experimental measurement data from the real system to the mathematical model using *system identification*, which in simple cases reduces to basic techniques like curve fitting and regression analysis. However, advanced versions of system identification may even determine the form of the mathematical model selected from a set of basic model structures.

1.5 ANALYZING MODELS

Simulation is one of the most common techniques for using models to answer questions about systems. However, there also exist other methods of analyzing models such as sensitivity analysis and model-based diagnosis or analytical mathematical techniques in the restricted cases where solutions can be found in a closed analytical form.

1.5.1 Sensitivity Analysis

Sensitivity analysis deals with the question how *sensitive* the behavior of the model is to *changes* of model parameters. This is a very common question in design and analysis of systems. For example, even in well-specified application domains such as electrical systems, resistor values in a circuit are typically known only by an accuracy of 5 to 10%. If there is a large sensitivity in the results of simulations to small variations in model parameters, we should be very suspicious about the validity the model. In such cases small random variations in the model parameters can lead to large random variations in the behavior.

On the other hand, if the simulated behavior is not very sensitive to small variations in the model parameters, there is a good chance that the model fairly accurately reflects the behavior of the real system. Such robustness in behavior is a desirable property when designing new products, since they otherwise may become expensive to manufacture since certain tolerances must be kept very small. However, there are also a number of examples of real systems which are very sensitive to variations of specific model parameters. In those cases that sensitivity should be reflected in models of those systems.

1.5.2 Model-Based Diagnosis

Model-based *diagnosis* is a technique somewhat related to sensitivity analysis. We want to find the causes of certain behavior of a system by analyzing a model of that system. In many cases we want to find the causes of problematic and erroneous behavior. For example, consider a car, which is a complex system consisting of many interacting parts such as a motor, an ignition system, a transmission system, suspension, wheels, and the like. Under a set of well-defined operating conditions each of these parts can be considered to exhibit a correct behavior if certain quantities are within specified value intervals. A measured or computed value outside such an interval might indicate an error in that component or in another part influencing that component. This kind of analysis is called model-based diagnosis.

1.5.3 Model Verification and Validation

We have previously remarked about the dangers of simulation, for example, when a model is not valid for a system regarding the intended simulation. How can we verify that the model is a good and reliable model, that is, is it valid for its intended use? This can be very hard, and sometimes we can hope only to get a partial answer to this question. However, the following techniques are useful to at least partially verify the validity of a model:

- Critically review the assumptions and approximations behind the model, including available information about the domain of validity regarding these assumptions.
- Compare simplified variants of the model to analytical solutions for special cases.
- Compare to experimental results for cases when this is possible.
- Perform sensitivity analysis of the model. If the simulation results are relatively insensitive to small variations of model parameters, we have stronger reasons to believe in the validity of the model.
- Perform internal consistency checking of the model, for example, checking that dimensions or units are compatible across equations. For example, in Newton's equation $F = ma$,

the unit (N) on the left-hand side is consistent with (kg m s^{-2}) on the right-hand side.

In the last case it is possible for tools to automatically verify that dimensions are consistent if unit attributes are available for the quantities of the model. This functionality, however, is not yet available for most current modeling tools.

1.6 KINDS OF MATHEMATICAL MODELS

Different kinds of mathematical models can be characterized by different properties reflecting the behavior of the systems that are modeled. One important aspect is whether the model incorporates *dynamic* time-dependent properties or is *static*. Another dividing line is between models that evolve *continuously* over time and those that change at *discrete* points in time. A third dividing line is between *quantitative* and *qualitative* models.

Certain models describe *physical distribution* of quantities, for example, mass, whereas other models are *lumped* in the sense that the physically distributed quantity is approximated by being lumped together and represented by a single variable, for example, a point mass.

Some phenomena in nature are conveniently described by stochastic processes and probability distributions, e.g. noisy radio transmissions or atomic-level quantum physics. Such models might be labeled *stochastic* or *probability*-based models where the behavior can be represented only in a statistic sense, whereas *deterministic* models allow the behavior to be represented without uncertainty. However, even stochastic models can be simulated in a "deterministic" way using a computer since the random number sequences often used to represent stochastic variables can be regenerated given the same seed values.

The same phenomenon can often be modeled as being either stochastic or deterministic depending on the level of detail at which it is studied. Certain aspects at one level are abstracted or averaged away at the next higher level. For example, consider the modeling of gases at different levels of detail starting at the quantum mechanical elementary particle level, where the positions of particles are described by probability distributions:

- Elementary particles (orbitals)—stochastic models
- Atoms (ideal gas model)—deterministic models
- Atom groups (statistical mechanics)—stochastic models
- Gas volumes (pressure and temperature)—deterministic models
- Real gases (turbulence)—stochastic models
- Ideal mixer (concentrations)—deterministic models

It is interesting to note the kinds of model changes between stochastic or deterministic models that occur depending on what aspects we want to study. Detailed stochastic models can be averaged as deterministic models when approximated at the next upper macroscopic level in the hierarchy. On the other hand, stochastic behavior such as turbulence can be introduced at macroscopic levels as the result of chaotic phenomena caused by interacting deterministic parts.

1.6.1 Kinds of Equations

Mathematical models usually contain equations. There are basically four main kinds of equations, where we give one example of each.

Differential equations contain time derivatives such as dx/dt, usually denoted $\dot{x}$, for example,

$$\dot{x} = a \cdot x + 3 \tag{1.1}$$

Algebraic equations do not include any differentiated variables:

$$x^2 + y^2 = L^2 \tag{1.2}$$

Partial differential equations also contain derivatives with respect to other variables than time:

$$\frac{\partial a}{\partial t} = \frac{\partial^2 a}{\partial z^2} \tag{1.3}$$

Difference equations express relations between variables, for example, at different points in time:

$$x(t+1) = 3x(t) + 2 \tag{1.4}$$

1.6.2 Dynamic Versus Static Models

All systems, both natural and man-made, are dynamic in the sense that they exist in the real world, which evolves in time. Mathematical models of such systems would be naturally viewed as *dynamic* in the sense that they evolve over time and therefore incorporate time. However, it is often useful to make the approximation of ignoring time dependence in a system. Such a system model is called *static*. Thus we can define the concepts of dynamic and static models as follows:

- A *dynamic* model includes *time* in the model. The word *dynamic* is derived from the Greek word *dynamis* meaning force and power, with dynamics being the (time-dependent) interplay between forces. Time can be included explicitly as a variable in a mathematical formula or be present indirectly, for example, through the time derivative of a variable or as events occurring at certain points in time.

- A *static* model can be defined *without* involving *time*, where the word *static* is derived from the Greek word *statikos*, meaning something that creates equilibrium. Static models are often used to describe systems in steady-state or equilibrium situations, where the output does not change if the input is the same. However, static models can display a rather dynamic behavior when fed with dynamic input signals.

It is usually the case that the behavior of a dynamic model is dependent on its *previous* simulation history. For example, the presence of a time derivative in a mathematical model means that this derivative needs to be integrated to solve for the corresponding variable when the model is simulated, that is, the *integration* operation takes the previous time history into account. This is the case, for example, for models of capacitors where the voltage over the capacitor is proportional to the accumulated charge in the capacitor, that is, integration/accumulation of the current through the capacitor. By differentiating that relation the time derivative of the capacitor voltage becomes proportional to the current through the capacitor. We can study the capacitor voltage increasing over time at a rate proportional to the current in Figure 1.3.

Another way for a model to be dependent on its previous history is to let preceding events influence the current state, for example, as in a

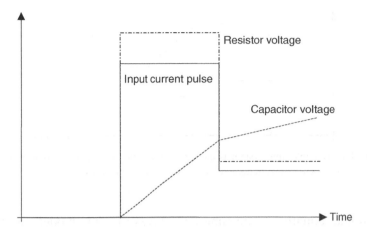

Figure 1.3 Resistor is a static system where the voltage is directly proportional to the current, independent of time, whereas a capacitor is a dynamic system where voltage is dependent on the previous time history.

model of an ecological system where the number of prey animals in the system will be influenced by events such as the birth of predators. On the other hand, a dynamic model such as a sinusoidal signal generator can be modeled by a formula directly including time and not involving the previous time history.

A resistor is an example of a static model that can be formulated without including time. The resistor voltage is directly proportional to the current through the resistor, for example, as depicted in Figure 1.3, with no dependence on time or on the previous history.

1.6.3 Continuous-Time Versus Discrete-Time Dynamic Models

There are two main classes of dynamic models: continuous-time and discrete-time models. The class of continuous-time models can be characterized as follows:

- *Continuous-time* models evolve their variable values continuously over time.

A variable from a continuous-time model A is depicted in Figure 1.4. The mathematical formulation of continuous-time models includes

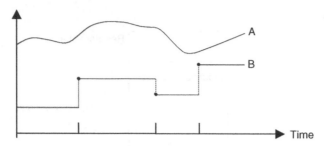

Figure 1.4 Discrete-time system B changes values only at certain points in time, whereas continuous-time systems like A evolve values continuously.

differential equations with time derivatives of some model variables. Many laws of nature, for example, as expressed in physics, are formulated as differential equations.

The second class of mathematical models is discrete-time models, for example, as B in Figure 1.4, where variables change value only at certain points in time:

- *Discrete-time* models may change their variable values only at discrete points in time.

Discrete-time models are often represented by sets of difference equations or as computer programs mapping the state of the model at one point in time to the state at the next point in time.

Discrete-time models occur frequently in engineering systems, especially computer-controlled systems. A common special case is sampled systems, where a continuous-time system is measured at regular time intervals and is approximated by a discrete-time model. Such sampled models usually interact with other discrete-time systems like computers. Discrete-time models may also occur naturally, for example, an insect population which breeds during a short period once a year; that is, the discretization period in that case is one year.

1.6.4 Quantitative Versus Qualitative Models

All of the different kinds of mathematical models previously discussed in this section are of a quantitative nature—variable values can be represented numerically according to a quantitatively measurable scale.

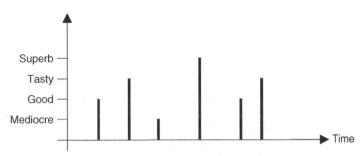

Figure 1.5   Quality of food in a restaurant according to inspections at irregular points in time.

Other models, so-called *qualitative* models, lack that kind of precision. The best we can hope for is a rough classification into a finite set of values, for example, as in the food quality model depicted in Figure 1.5. Qualitative models are by nature discrete-time models, and the dependent variables are also discretized. However, even if the discrete values are represented by numbers in the computer (e.g., mediocre—1, good—2, tasty—3, superb—4), we have to be aware of the fact that the values of variables in certain qualitative models are not necessarily according to a linear measurable scale, that is, tasty might not be three times better than mediocre.

1.7 USING MODELING AND SIMULATION IN PRODUCT DESIGN

What role does modeling and simulation have in industrial product design and development? In fact, our previous discussion has already briefly touched this issue. Building mathematical models in the computer, so-called *virtual prototypes*, and simulating those models, is a way to quickly determine and optimize product properties without building costly physical prototypes. Such an approach can often drastically reduce development time and time to market, while increasing the quality of the designed product.

The so-called product *design V*, depicted in Figure 1.6, includes all the standard phases of product development:

- Requirements analysis and specification
- System design

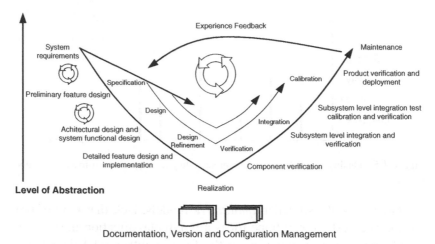

Figure 1.6 Product design V.

- Design refinement
- Realization and implementation
- Subsystem verification and validation
- Integration
- System calibration and model validation
- Product deployment

How does modeling and simulation fit into this design process?

In the first phase, *requirements analysis*, functional and nonfunctional requirements are specified. In this phase important design parameters are identified and requirements on their values are specified. For example, when designing a car, there might be requirements on acceleration, fuel consumption, maximum emissions, and the like. Those system parameters will also become parameters in our model of the designed product.

In the *system design phase* we specify the architecture of the system, that is, the main components in the system and their interactions. If we have a simulation model component library at hand, we can use these library components in the design phase or otherwise create new components that fit the designed product. This design process iteratively increases the level of detail in the design. A modeling tool that supports hierarchical system modeling and decomposition can help in handling system complexity.

The *implementation phase* will realize the product as a physical system and/or as a virtual prototype model in the computer. Here a virtual prototype can be realized before the physical prototype is built, usually for a small fraction of the cost.

In the *subsystem verification and validation phase*, the behavior of the subsystems of the product is verified. The subsystem virtual prototypes can be simulated in the computer and the models corrected if there are problems.

In the *integration phase* the subsystems are connected. Regarding a computer-based system model, the models of the subsystems are connected together in an appropriate way. The whole system can then be simulated, and certain design problems corrected based on the simulation results.

The system and model *calibration and validation phase* validates the model against measurements from appropriate physical prototypes. Design parameters are calibrated, and the design is often *optimized* to a certain extent according to what is specified in the original requirements.

During the last phase, *product deployment*, which usually only applies to the physical version of the product, the product is deployed and sent to the customer for feedback. In certain cases this can also be applied to virtual prototypes, which can be delivered and put in a computer that is interacting with the rest of the customer physical system in real time, that is, hardware-in-the-loop simulation.

In most cases, experience feedback can be used to tune both models and physical products. All phases of the design process continuously interact with the model and design database, as depicted at the bottom of Figure 1.6.

1.8 EXAMPLES OF SYSTEM MODELS

In this section we briefly present examples of mathematical models from three different application areas, in order to illustrate the power of the Modelica mathematical modeling and simulation technology to be described in the rest of this book:

- A thermodynamic system—part of an industrial GTX100 gas turbine model
- A three-dimensional (3D) mechanical system with a hierarchical decomposition—an industry robot
- A biochemical application—part of the citrate cycle (TCA cycle), see Figure 1.11

A connection diagram of the power cutoff mechanism of the GTX100 gas turbine is depicted in Figure 1.8, whereas the gas turbine itself is shown in Figure 1.7.

The connection diagram in Figure 1.8 might not appear as a mathematical model, but behind each icon in the diagram is a model component containing the equations that describe the behavior of the respective component.

In Figure 1.9 we show a few plots from simulations of the gas turbine, which illustrates how a model can be used to investigate the properties of a given system.

The second example, the industry robot, illustrates the power of hierarchical model decomposition. The 3D robot, shown to the right of Figure 1.10, is represented by a two-dimensional (2D) connection diagram (in the middle). Each part in the connection diagram can be a mechanical component such as a motor or joint, a control system for the robot, and so forth. Components may consist of other components that can in turn be decomposed. At the bottom of the hierarchy wehave model classes containing the actual equations.

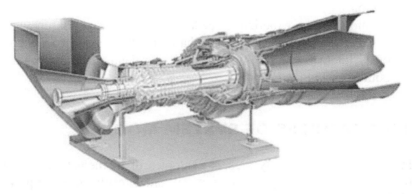

Figure 1.7 Schematic picture of the gas turbine GTX100. (Courtesy Siemens Industrial Turbomachinery AB, Finspång, Sweden.)

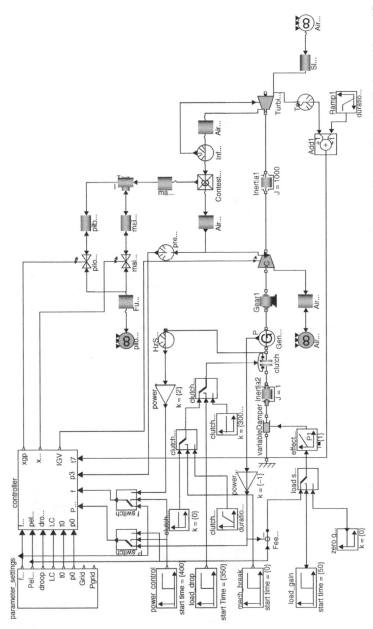

Figure 1.8 Detail of power cutoff mechanism in 40 MW GTX100 gas turbine model. (Courtesy Siemens Industrial Turbomachinery AB, Finspång, Sweden.)

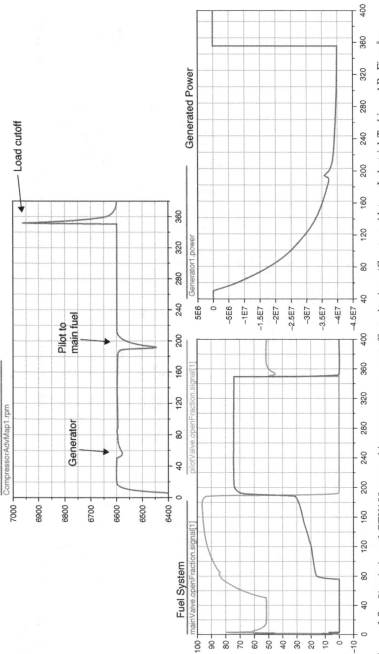

Figure 1.9 Simulation of GTX100 gas turbine power system cutoff mechanism. (Courtesy Alstom Industrial Turbines AB, Finspång, Sweden.)

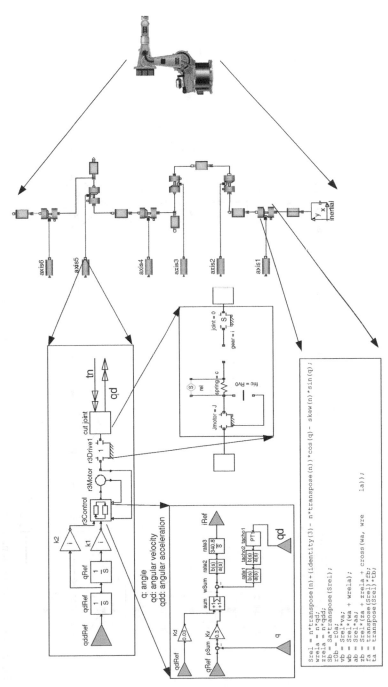

Figure 1.10 Hierarchical model of an industrial robot. (Courtesy Martin Otter.)

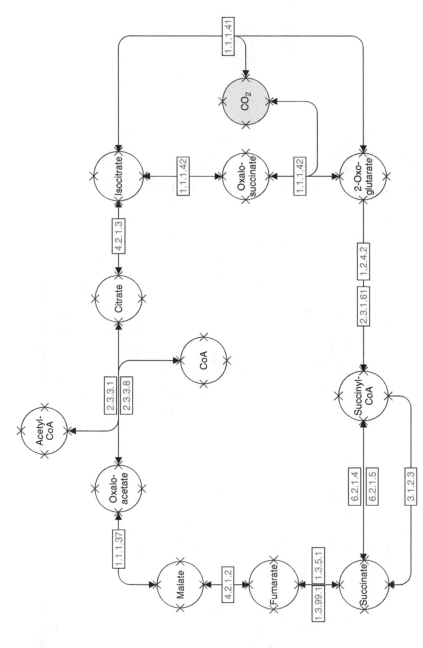

Figure 1.11 Biochemical pathway model of part of the citrate cycle (TCA cycle).

The third example is from an entirely different domain—biochemical pathways describing the reactions between reactants, in this particular case describing part of the citrate cycle (TCA cycle) as depicted in Figure 1.11.

1.9 SUMMARY

We have briefly presented important concepts such as system, model, experiment, and simulation. Systems can be represented by models, which can be subject to experiments, that is, simulation. Certain models can be represented by mathematics, so-called mathematical models. This book is about object-oriented component-based technology for building and simulating such mathematical models. There are different classes of mathematical models, for example, static versus dynamic models, continuous-time versus discrete-time models, and so forth, depending on the properties of the modeled system, the available information about the system, and the approximations made in the model.

1.10 LITERATURE

Any book on modeling and simulation needs to define fundamental concepts such as system, model, and experiment. The definitions in this chapter are generally available in modeling and simulation literature, including Ljung and Glad (1994) and Cellier (1991). The example of different levels of details in mathematical models of gases presented in Section 1.6 is mentioned in Hyötyniemi (2002). The product design-V process mentioned in Section 1.7 is described in Stevens et al. (1998) and Shumate and Keller (1992). The citrate cycle biochemical pathway part in Figure 1.11 is modeled after the description in Allaby (1998).

CHAPTER 2

A Quick Tour of Modelica

Modelica is primarily a modeling language that allows specification of mathematical models of complex natural or man-made systems, for example, for the purpose of computer simulation of dynamic systems where behavior evolves as a function of time. Modelica is also an object-oriented equation-based programming language, oriented toward computational applications with high complexity requiring high performance. The four most important features of Modelica are:

- Modelica is primarily based on equations instead of assignment statements. This permits acausal modeling that gives better reuse of classes since equations do not specify a certain data flow direction. Thus a Modelica class can adapt to more than one data flow context.

- Modelica has multidomain modeling capability, meaning that model components corresponding to physical objects from several different domains such as, for example, electrical, mechanical, thermodynamic, hydraulic, biological, and control applications can be described and connected.

- Modelica is an object-oriented language with a general class concept that unifies classes, generics—known as templates in C++—and general subtyping into a single language construct. This facilitates reuse of components and evolution of models.

- Modelica has a strong software component model, with constructs for creating and connecting components. Thus

Introduction to Modeling and Simulation of Technical and Physical Systems with Modelica,
First Edition. By Peter Fritzson
© 2011 the Institute of Electrical and Electronics Engineers, Inc. Published 2011 by John Wiley & Sons, Inc.

the language is ideally suited as an architectural description language for complex physical systems and to some extent for software systems.

These are the main properties that make Modelica both powerful and easy to use, especially for modeling and simulation. We will start with a gentle introduction to Modelica from the very beginning.

2.1 GETTING STARTED WITH MODELICA

Modelica programs are built from classes, also called models. From a class definition, it is possible to create any number of objects that are known as instances of that class. Think of a class as a collection of blueprints and instructions used by a factory to create objects. In this case the Modelica compiler and run-time system is the factory.

A Modelica class contains elements, the main kind being variable declarations, and equation sections containing equations. Variables contain data belonging to instances of the class; they make up the data storage of the instance. The equations of a class specify the behavior of instances of that class.

There is a long tradition that the first sample program in any computer language is a trivial program printing the string "Hello World". Since Modelica is an equation-based language, printing a string does not make much sense. Instead, our Hello World Modelica program solves a trivial *differential equation*:

$$\dot{x} = -a \cdot x \tag{2.1}$$

The variable x in this equation is a dynamic variable (here also a state variable) that can change value over time. The time derivative $\dot{x}$ is the time derivative of x, represented as der(x) in Modelica. Since all Modelica programs, usually called *models*, consist of class declarations, our HelloWorld program is declared as a class:

```
class HelloWorld
  Real x(start = 1);
  parameter Real a = 1;
equation
  der(x) = -a*x;
end HelloWorld;
```

Use your favorite text editor or Modelica programming environment to type in this Modelica code,[1] or open the DrModelica electronic document containing all examples and exercises in this book. Then invoke the simulation command in your Modelica environment. This will compile the Modelica code to some intermediate code, usually C code, which in turn will be compiled to machine code and executed together with a numerical ordinary differential equation (ODE) solver or differential algebraic equation (DAE) solver to produce a solution for x as a function of time. The following command in the OpenModelica environment produces a solution between time 0 and time 2:

```
simulate² (HelloWorld, stopTime=2)
```

Since the solution for x is a function of time, it can be plotted by a plot command:

```
plot³(x)
```

(or the longer form `plot(x,xrange={0,2})` that specifies the x axis), giving the curve in Figure 2.1.

Now we have a small Modelica model that does something, but what does it actually mean? The program contains a declaration of a class called `HelloWorld` with two variables and a single equation. The first attribute of the class is the variable x, which is initialized to a start value of 1 at the time when the simulation starts. All variables in Modelica have a `start` attribute with a default value that is normally set to 0. Having a different start value is accomplished by providing a so-called modifier within parentheses after the variable name, that is, a modification equation setting the start attribute to 1 and replacing the original default equation for the attribute.

---

[1]There is an open-source Modelica environment OpenModelica downloadable from www.openmodelica.org, MathModelica is from Wolfram Research and MathCore (www.mathcore.com), and Dymola from Dassault Systémès, (www.3ds.com/products/catia/portfolio/dymola).

[2]`simulate` is the OpenModelica command for simulation. The corresponding MathModelica Mathematica-style command for this example would be `Simulate[HelloWorld, {t,0,2}]`, and in Dymola `simulateModel("HelloWorld", stopTime=2)`.

[3]`plot` is the OpenModelica command for plotting simulation results. The corresponding MathModelica Mathematica-style and Dymola commands would be `PlotSimulation[x[t], {t,0,2}]` and `plot({"x"})`, respectively.

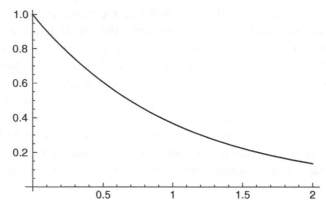

Figure 2.1   Plot of simulation of simple `HelloWorld` model.

The second attribute is the variable a, which is a constant that is initialized to 1 at the beginning of the simulation. Such a constant is prefixed by the keyword `parameter` in order to indicate that it is constant during simulation but is a model parameter that can be changed between simulations, for example, through a command in the simulation environment. For example, we could rerun the simulation for a different value of a.

Also note that each variable has a type that precedes its name when the variable is declared. In this case both the variable x and the "variable" a have the type `Real`.

The single equation in this `HelloWorld` example specifies that the time derivative of x is equal to the constant -a times x. In Modelica the equal sign = always means equality, that is, establishes an equation, and not an assignment as in many other languages. Time derivative of a variable is indicated by the pseudofunction `der( )`.

Our second example is only slightly more complicated, containing five rather simple equations:

$$m\dot{v}_x = -\frac{x}{L}F$$
$$m\dot{v}_y = -\frac{y}{L}F - mg \qquad (2.2)$$
$$\dot{x} = v_x$$
$$\dot{y} = v_y$$
$$x^2 + y^2 = L^2$$

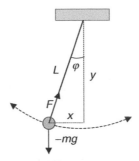

Figure 2.2   Planar pendulum.

This example is actually a mathematical model of a physical system, a planar pendulum, as depicted in Figure 2.2. The equations are Newton's equations of motion for the pendulum mass under the influence of gravity, together with a geometric constraint, the fifth equation, $x^2 + y^2 = L^2$, which specifies that its position (x, y) must be on a circle with radius L. The variables v_x and v_y are its velocities in the x and y directions, respectively.

The interesting property of this model, however, is the fact that the fifth equation is of a different kind: a so-called *algebraic equation* only involving algebraic formulas of variables but no derivatives. The first four equations of this model are differential equations as in the `HelloWorld` example. Equation systems that contain both differential and algebraic equations are called *differential algebraic equation systems* (DAEs). A Modelica model of the pendulum appears below:

```
class Pendulum   "Planar Pendulum"
   constant   Real PI=3.141592653589793;
   parameter Real m=1, g=9.81, L=0.5;
   Real F;
   output    Real x(start=0.5),y(start=0);
   output    Real vx,vy;
equation
  m*der(vx)=-(x/L)*F;
  m*der(vy)=-(y/L)*F-m*g;
  der(x)=vx;
  der(y)=vy;
  x^2+y^2=L^2;
end Pendulum;
```

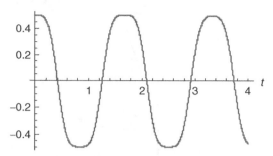

Figure 2.3 Plot of simulation of Pendulum DAE (differential algebraic equation) model.

We simulate the Pendulum model and plot the x coordinate, shown in Figure 2.3:

```
simulate(Pendulum, stopTime=4)
plot(x);
```

You can also write down DAE systems without physical significance, with equations containing formulas selected more or less at random, as in the class DAEexample below:

```
class DAEexample
  Real x(start=0.9);
  Real y;
equation
  der(y) + (1+0.5*sin(y))*der(x) = sin(time);
  x-y = exp(-0.9*x)*cos(y);
end DAEexample;
```

This class contains one differential and one algebraic equation. Try to simulate and plot it yourself, to see if any reasonable curve appears! Finally, an important observation regarding Modelica models:

- The number of *variables* must be equal to the number of *equations*!

This statement is true for the three models we have seen so far and holds for all solvable Modelica models. By variables we here mean something that can vary, that is, not named constants and parameters already having values, described in Section 2.1.3.

2.1.1 Variables and Predefined Types

This example shows a slightly more complicated model, which describes a van der Pol[4] oscillator. Notice that here the keyword model is used instead of class with almost the same meaning.

```
model VanDerPol    "Van der Pol oscillator model"
  Real x(start = 1)   "Descriptive string for x";
    // x starts at 1
  Real y(start = 1)   "Descriptive string for y";
    // y starts at 1
  parameter Real lambda = 0.3;
equation
  der(x) = y;
    // This is the first equation
  der(y) = -x + lambda*(1 - x*x)*y;
    /* The 2nd diff. equation */
end VanDerPol;
```

This example contains declarations of two dynamic variables (here also state variables) x and y, both of type Real and having the start value 1 at the beginning of the simulation, which normally is at time 0. Then follows a declaration of the parameter constant lambda, which is a so-called model parameter.

The keyword parameter specifies that the variable is constant during a simulation run but can have its value initialized before a run or between runs. This means that parameter is a special kind of constant, which is implemented as a static variable that is initialized once and never changes its value during a specific execution. A parameter is a constant variable that makes it simple for a user to modify the behavior of a model, for example, changing the parameter lambda, which strongly influences the behavior of the Van der Pol oscillator. By contrast, a fixed Modelica constant declared with the prefix constant never changes and can be substituted by its value wherever it occurs.

Finally, we present declarations of three dummy variables just to show variables of data types different from Real: the Boolean variable bb, which has a default start value of false if nothing

---

[4]Balthazar van der Pol was a Dutch electrical engineer who initiated modern experimental dynamics in the laboratory during the 1920s and 1930s. Van der Pol investigated electrical circuits employing vacuum tubes and found that they have stable oscillations, now called limit cycles. The van der Pol oscillator is a model developed by him to describe the behavior of nonlinear vacuum tube circuits.

else is specified, the string variable `dummy` which is always equal to "`dummy string`", and the integer variable `fooint` always equal to 0.

```
Boolean bb;
String dummy = "dummy string";
Integer fooint = 0;
```

Modelica has built-in predefined "primitive" data types to support floating-point, integer, Boolean, and string values. There is also the `Complex` type for complex numbers computations, which is predefined in a library. These predefined types contain data that Modelica understands directly, as opposed to class types defined by programmers. The type of each variable must be declared explicitly. The predefined basic data types of Modelica are:

`Boolean`	either `true` or `false`
`Integer`	corresponding to the C int data type, usually 32-bit two's complement
`Real`	corresponding to the C double data type, usually 64-bit floating-point
`String`	string of text characters
`enumeration(...)`	enumeration type of enumeration literals
`Complex`	for complex number computations, a basic type predefined in a library

Finally, there is an equation section starting with the keyword `equation`, containing two mutually dependent equations that define the dynamics of the model.

To illustrate the behavior of the model, we give a command to simulate the van der Pol oscillator during 25 s starting at time 0:

```
simulate(VanDerPol, stopTime=25)
```

A phase plane plot of the state variables for the van der Pol oscillator model (Fig. 2.4):

```
plotParametric(x,y, stopTime=25)
```

The names of variables, functions, classes, and so forth are known as identifiers. There are two forms in Modelica. The most common form starts with a letter, followed by letters or digits, for example, `x2`. The second form starts with a single quote, followed by any characters, and terminated by a single quote, for example, `'2nd*3'`.

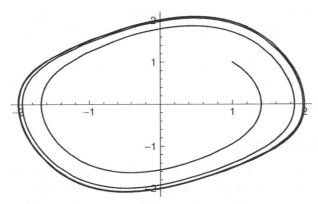

Figure 2.4   Parametric plot of simulation of van der Pol oscillator model.

2.1.2 Comments

Arbitrary descriptive text, for example, in English, inserted through-out a computer program are comments to that code. Modelica has three styles of comments, all illustrated in the previous `VanDerPol` example.

Comments make it possible to write descriptive text together with the code, which makes a model easier to use for the user, or easier to understand for programmers who may read your code in the future. That programmer may very well be yourself, months or years later. You save yourself future effort by commenting your own code. Also, it is often the case that you find errors in your code when you write comments since when explaining your code you are forced to think about it once more.

The first kind of comment is a string within string quotes, for example, "a `comment`," optionally appearing after variable declara-tions or at the beginning of class declarations. Those are "definition comments" that are processed to be used by the Modelica program-ming environment, for example, to appear in menus or as help texts for the user. From a syntactic point of view they are not really comments since they are part of the language syntax. In the previous example such definition comments appear for the `VanDerPol` class and for the x and y variables.

The other two types of comments are ignored by the Modelica compiler and are just present for the benefit of Modelica programmers. Text following `//` up to the end of the line is skipped by the compiler, as is text between /* and the next */. Hence the last type of comment can be used for large sections of text that occupies several lines.

Finally we should mention a construct called annotation, a kind of structured "comment" that can store information together with the code, described in Section 2.17.

2.1.3 Constants

Constant literals in Modelica can be integer values such as 4,75,3078; floating-point values like 3.14159,0.5,2.735E-10, 8.6835e+5; string values such as "hello world", "red"; and enumeration values such as Colors.red,Sizes.xlarge.

Named constants are preferred by programmers for two reasons. One reason is that the name of the constant is a kind of documentation that can be used to describe what the particular value is used for. The other, perhaps even more important reason, is that a named constant is defined at a single place in the program. When the constant needs to be changed or corrected, it can be changed in only one place, simplifying program maintenance.

Named constants in Modelica are created by using one of the prefixes constant or parameter in declarations, and providing a *declaration equation* as part of the declaration:

```
constant   Real     PI = 3.141592653589793;
constant   String   redcolor = "red";
constant   Integer  one = 1;
parameter  Real     mass = 22.5;
```

Parameter constants can be declared without a declaration equation since their value can be defined, for example, by reading from a file, before simulation starts. For example:

```
parameter Real mass, gravity, length;
```

2.1.4 Variability

We have seen that some variables can change value at any point in time, whereas named constants are more or less constant. In fact, there is a general concept of four levels of variability of variables and expressions in Modelica:

- Expressions or variables with *continuous-time variability* can change at any point in time.

- *Discrete-time variability* means that value changes can occur only at so-called events; see Section 2.15.

- *Parameter variability* means that the value can be changed at initialization before simulation but is fixed during simulation.

- *Constant variability* means the value is always fixed.

2.1.5 Default `start` Values

If a numeric variable lacks a specified definition value or `start` value in its declaration, it is usually initialized to zero at the start of the simulation. Boolean variables have `start` value `false`, and string variables have the `start` value empty string `" "` if nothing else is specified.

Exceptions to this rule are function *results* and *local* variables in functions, where the default initial value at function call is *undefined*.

2.2 OBJECT-ORIENTED MATHEMATICAL MODELING

Traditional object-oriented programming languages like Simula, C++, Java, and Smalltalk, as well as procedural languages such as Fortran or C, support programming with operations on stored data. The stored data of the program include variable values and object data. The number of objects often changes dynamically. The Smalltalk view of object orientation emphasizes sending messages between (dynamically) created objects.

The Modelica view on object orientation is different since the Modelica language emphasizes *structured* mathematical modeling. Object orientation is viewed as a structuring concept that is used to handle the complexity of large system descriptions. A Modelica model is primarily a declarative mathematical description, which simplifies further analysis. Dynamic system properties are expressed in a declarative way through equations.

The concept of *declarative* programming is inspired by mathematics, where it is common to state or declare what *holds*, rather than giving a detailed stepwise *algorithm* on *how* to achieve the desired goal as is required when using procedural languages. This relieves the programmer from the burden of keeping track of such details. Furthermore, the code becomes more concise and easier to change without introducing errors.

Thus, the declarative Modelica view of object orientation, from the point of view of object-oriented mathematical modeling, can be summarized as follows:

- Object orientation is primarily used as a *structuring* concept, emphasizing the declarative structure and reuse of mathematical models. Our three ways of structuring are hierarchies, component connections, and inheritance.
- Dynamic model properties are expressed in a declarative way through *equations*.[5]
- An object is a collection of *instance* variables and equations that share a set of data.

However:

- Object orientation in mathematical modeling is *not* viewed as dynamic message passing.

The declarative object-oriented way of describing systems and their behavior offered by Modelica is at a higher level of abstraction than the usual object-oriented programming since some implementation details can be omitted. For example, we do not need to write code to explicitly transport data between objects through assignment statements or message-passing code. Such code is generated automatically by the Modelica compiler based on the given equations.

Just as in ordinary object-oriented languages, classes are blueprints for creating objects. Both variables and equations can be inherited between classes. Function definitions can also be inherited. However, specifying behavior is primarily done through equations instead of via methods. There are also facilities for stating algorithmic code including functions in Modelica, but this is an exception rather than the rule. See also Chapter 3 for a discussion regarding object-oriented concepts.

---

[5]Algorithms are also allowed, but in a way that makes it possible to regard an algorithm section as a system of equations.

2.3 CLASSES AND INSTANCES

Modelica, like any object-oriented computer language, provides the notions of classes and objects, also called instances, as a tool for solving modeling and programming problems. Every object in Modelica has a class that defines its data and behavior. A class has three kinds of members:

- Data variables associated with a class and its instances. Variables represent results of computations caused by solving the equations of a class together with equations from other classes. During numeric solution of time-dependent problems, the variables store results of the solution process at the current time instant
- Equations specify the behavior of a class. The way in which the equations interact with equations from other classes determines the solution process, that is, program execution.
- Classes can be members of other classes.

Here is the declaration of a simple class that might represent a point in a three-dimensional space:

```
class Point "Point in a three-dimensional space"
  public
    Real x;
    Real y, z;
  end Point;
```

The `Point` class has three variables representing the x, y, and z coordinates of a point and has no equations. A class declaration like this one is like a blueprint that defines how instances created from that class look like, as well as instructions in the form of equations that define the behavior of those objects. Members of a class may be accessed using dot (.) notation. For example, regarding an instance `myPoint` of the `Point` class, we can access the x variable by writing `myPoint.x`.

Members of a class can have two levels of visibility. The `public` declaration of x, y, and z, which is default if nothing else is specified, means that any code with access to a `Point` instance can refer to those values. The other possible level of visibility, specified by the keyword `protected`, means that only code inside the class as well as code in classes that inherit this class, are allowed access.

Note that an occurrence of one of the keywords `public` or `protected` means that all member declarations following that keyword assume the corresponding visibility until another occurrence of one of those keywords, or the end of the class containing the member declarations has been reached.

2.3.1 Creating Instances

In Modelica, objects are created implicitly just by declaring instances of classes. This is in contrast to object-oriented languages like Java or C++, where object creation is specified using the new keyword. For example, to create three instances of our `Point` class we just declare three variables of type `Point` in a class, here `Triangle`:

```
class Triangle
   Point    point1;
   Point    point2;
   Point    point3;
end Triangle;
```

There is one remaining problem, however. In what context should `Triangle` be instantiated, and when should it just be interpreted as a library class not to be instantiated until actually used?

This problem is solved by regarding the class at the *top* of the instantiation hierarchy in the Modelica program to be executed as a kind of "main" class that is always implicitly instantiated, implying that its variables are instantiated, and that the variables of those variables are instantiated, and so forth. Therefore, to instantiate `Triangle`, either make the class `Triangle` the "top" class or declare an instance of `Triangle` in the "main" class. In the following example, both the class `Triangle` and the class `Foo1` are instantiated:

```
class Foo1
   ...
end Foo1;

class Foo2
   ...
end Foo2;
```

. . .

```
class Triangle
   Point   point1;
   Point   point2;
   Point   point3;
end Triangle;

class Main
   Triangle   pts;
   Foo1       f1;
end Main;
```

The variables of Modelica classes are instantiated per object. This means that a variable in one object is distinct from the variable with the same name in every other object instantiated from that class. Many object-oriented languages allow class variables. Such variables are specific to a class as opposed to instances of the class, and are shared among all objects of that class. The notion of class variables is not yet available in Modelica.

2.3.2 Initialization

Another problem is initialization of variables. As mentioned previously in Section 2.1.5, if nothing else is specified, the default start value of all numerical variables is zero, apart from function results and local variables where the initial value at call time is unspecified. Other start values can be specified by setting the start attribute of instance variables. Note that the start value only gives a suggestion for initial value—the solver may choose a different value unless the fixed attribute is true for that variable. Below a start value is specified in the example class Triangle:

```
class Triangle
   Point   point1(start={1,2,3});
   Point   point2;
   Point   point3;
end Triangle;
```

Alternatively, the start value of point1 can be specified when instantiating Triangle as below:

```
class Main
   Triangle   pts(point1.start={1,2,3});
```

```
    foo1        f1;
end Main;
```

A more general way of initializing a set of variables according to some constraints is to specify an equation system to be solved in order to obtain the initial values of these variables. This method is supported in Modelica through the `initial equation` construct.

An example of a continuous-time controller initialized in steady state, that is, when derivatives should be zero, is given below:

```
model Controller
  Real y;
equation
  der(y) = a*y + b*u;
initial equation
  der(y)=0;
end Controller;
```

This has the following solution at initialization:

```
der(y) = 0;
y = -(b/a)*u;
```

2.3.3 Specialized Classes

The class concept is fundamental to Modelica and is used for a number of different purposes. Almost anything in Modelica is a class. However, in order to make Modelica code easier to read and maintain, special keywords have been introduced for specific uses of the class concept. The keywords `model`, `connector`, `record`, `block`, `type`, `package`, and `function` can be used to denote a class under appropriate conditions, called restrictions. Some of the specialized classes also have additional capabilities, called enhancements. For example, a `function` class has the enhancement that it can be called, whereas a `record` is a class used to declare a record data structure and has the restriction that it may not contain equations.

```
record Person
  Real    age;
  String name;
end Person;
```

A `model` is the same as a `class`, that is, those keywords are completely interchangeable. A `block` is a class with fixed causality, which means that for each member variable of the class it is specified whether it has input or output causality. Thus, each variable in a block class interface must be declared with a causality prefix keyword of either input or output.

A connector class is used to declare the structure of "ports" or interface points of a component and may not contain equations, but has the additional property to allow connect(..) to instances of connector classes. A type is a class that can be an alias or an extension to a predefined type, record, or array. For example:

```
type vector3D = Real[3];
```

The idea of specialized classes is beneficial since the user does not have to learn several different concepts, except for one: the *class concept*. The notion of specialized classes gives the user a chance to express more precisely what a class is intended for and requires the Modelica compiler to check that these usage constraints are actually fulfilled. Fortunately, the notion is quite uniform since all basic properties of a class, such as the syntax and semantics of definition, instantiation, inheritance, and generic properties, are identical for all kinds of specialized classes. Furthermore, the construction of Modelica translators is simplified because only the syntax and semantics of the class concept have to be implemented along with some additional checks on specialized classes.

The package and function specialized classes in Modelica have much in common with the class concept but also have additional properties, so-called enhancements. Especially functions have quite a lot of enhancements, for example, it can be called with an argument list, instantiated at run time, and so forth. An operator class is similar to a package but may only contain declarations of functions and is intended for user-defined overloaded operators (Section 2.14.4).

2.3.4 Reuse of Classes by Modifications

The class concept is the key to reuse of modeling knowledge in Modelica. Provisions for expressing adaptations or *modifications* of classes through so-called modifiers in Modelica make reuse easier. For example, assume that we would like to connect two filter models with different time constants in series.

Instead of creating two separate filter classes, it is better to define a common filter class and create two appropriately modified instances of this class, which are connected. An example of connecting two modified low-pass filters is shown after the example low-pass filter class below:

```
model LowPassFilter
  parameter Real T=1  "Time constant of filter";
  Real u, y(start=1);
equation
  T*der(y) + y = u;
end LowPassFilter;
```

The model class can be used to create two instances of the filter with different time constants and "connecting" them together by the equation F2.u = F1.y as follows:

```
model FiltersInSeries
  LowPassFilter F1(T=2), F2(T=3);
equation
  F1.u = sin(time);
  F2.u = F1.y;
end FiltersInSeries;
```

Here we have used *modifiers*, that is, attribute equations such as T=2 and T=3 to modify the time constant of the low-pass filter when creating the instances F1 and F2. The independent time variable is denoted time. If the FiltersInSeries model is used to declare variables at a higher hierarchical level, for example, F12, the time constants can still be adapted by using hierarchical modification, as for F1 and F2 below:

```
model ModifiedFiltersInSeries
  FiltersInSeries F12(F1(T=6), F2.T=11);
end ModifiedFiltersInSeries;
```

2.3.5 Built-in Classes and Attributes

The built-in type classes of Modelica correspond to the predefined "primitive" types Real, Integer, Boolean, String, and enumeration(...), and have most of the properties of a class, for example, can be inherited, modified, and so forth. Only the value attribute can be changed at run time and is accessed through the variable name itself, and not through dot notation, that is, use x and not x.value to access the value. Other attributes are accessed through dot notation.

For example, a Real variable has a set of default attributes such as unit of measure, initial value, and minimum and maximum values. These default attributes can be changed when declaring a new class, for example:

```
class Voltage = Real(unit= "V", min=-220.0,
   max=220.0);
```

2.4 INHERITANCE

One of the major benefits of object orientation is the ability to extend the behavior and properties of an existing class. The original class, known as the *superclass* or *base class*, is extended to create a more specialized version of that class, known as the *subclass* or *derived class*. In this process, the behavior and properties of the original class in the form of variable declarations, equations, and other contents are reused, or inherited, by the subclass.

Let us regard an example of extending a simple Modelica class, for example, the class Point introduced previously. First, we introduce two classes named ColorData and Color, where Color inherits the data variables to represent the color from class ColorData and adds an equation as a constraint. The new class ColoredPoint inherits from multiple classes, that is, uses multiple inheritance, to get the position variables from class Point, and the color variables together with the equation from class Color.

```
record ColorData
   Real red;
   Real blue;
   Real green;
end ColorData;

class Color
   extends ColorData;
equation
   red + blue + green = 1;
end Color;
```

```
class Point
 public
  Real x;
  Real y, z;
end Point;

class ColoredPoint
   extends Point;
   extends Color;
 end ColoredPoint;
```

See also Section 3.7 regarding inheritance and reuse.

2.5 GENERIC CLASSES

In many situations it is advantageous to be able to express generic patterns for models or programs. Instead of writing many similar pieces of code with essentially the same structure, a substantial amount of coding and software maintenance can be avoided by directly expressing the general structure of the problem and providing the special cases as *parameter* values.

Such generic constructs are available in several programming languages, for example, templates in C++, generics in Ada, and type parameters in functional languages such as Haskell or Standard ML. In Modelica the class construct is sufficiently general to handle generic modeling and programming in addition to the usual class functionality.

There are essentially two cases of generic class parameterization in Modelica: *Class parameters* can either be *instance parameters*, that is, have instance declarations (components) as values, or be *type parameters*, that is, have types as values. Note that by class parameters in this context we do not usually mean model parameters prefixed by the keyword `parameter`, even though such "variables" are also a kind of class parameter. Instead we mean *formal parameters to the class*. Such formal parameters are prefixed by the keyword `replaceable`. The special case of replaceable local functions is roughly equivalent to virtual methods in some object-oriented programming languages.

2.5.1 Class Parameters as Instances

First, we present the case when class parameters are variables, that is, declarations of instances, often called components. The class C in the

example below has three class parameters *marked* by the keyword
`replaceable`. These class parameters, which are components
(variables) of class C, are declared as having the (default) types
`GreenClass`, `YellowClass`, and `GreenClass`, respectively. There
is also a red object declaration that is not replaceable and therefore
not a class parameter (Fig. 2.5).

Here is the `class` C with its three class parameters `pobj1`,
`pobj2`, and `pobj3` and a variable `obj4` that is not a class parameter:

```
class C
  replaceable GreenClass  pobj1(p1=5);
  replaceable YellowClass pobj2;
  replaceable GreenClass  pobj3;
  RedClass    obj4;
equation
  . . .
end C;
```

Now a class C2 is defined by providing two declarations of `pobj1`
and `pobj2` as actual arguments to class C, being `red` and `green`,
respectively, instead of the defaults `green` and `yellow`. The keyword `redeclare` must precede an actual argument to a class formal
parameter to allow changing its type. The requirement to use a keyword for a redeclaration in Modelica has been introduced in order to
avoid accidentally changing the type of an object through a standard
modifier.

In general, the type of a class component cannot be changed if it
is not declared as `replaceable` and a redeclaration is provided. A
variable in a redeclaration can replace the original variable if it has
a type that is a subtype of the original type or its type constraint. It
is also possible to declare type constraints (not shown here) on the
substituted classes.

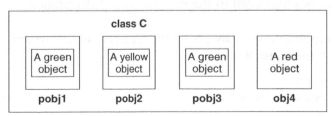

Figure 2.5 Three class parameters `pobj1`, `pobj2`, and `pobj3` that are instances (variables) of `class` C. These are essentially slots that can contain objects of different colors.

```
class C2 = C(redeclare RedClass pobj1, redeclare
   GreenClass pobj2);
```

Such a class C2 obtained through redeclaration of pobj1 and pobj2 is of course equivalent to directly defining C2 without reusing class C, as below.

```
class C2
   RedClass    pobj1(p1=5);
   GreenClass  pobj2;
   GreenClass  pobj3;
   RedClass    obj4;
equation
   ...
end C2;
```

2.5.2 Class Parameters as Types

A class parameter can also be a type, which is useful for changing the type of many objects. For example, by providing a type parameter ColoredClass in class C below, it is easy to change the color of all objects of type ColoredClass.

```
class C
   replaceable class ColoredClass = GreenClass;
   ColoredClass            obj1(p1=5);
   replaceable YellowClass obj2;
   ColoredClass            obj3;
   RedClass                obj4;
equation
   ...
end C;
```

Figure 2.6 depicts how the type value of the ColoredClass class parameter is propagated to the member object declarations obj1 and obj3.

We create a class C2 by giving the type parameter ColoredClass of class C the value BlueClass:

```
class C2 =
   C(redeclare class ColoredClass = BlueClass);
```

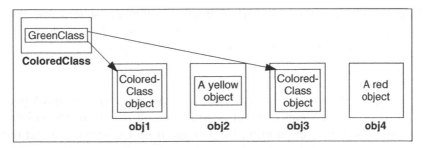

Figure 2.6   Class parameter `ColoredClass` is a type parameter that is propagated to the two-member instance declarations of `obj1` and obj3.

This is equivalent to the following definition of C2:

```
class C2
   BlueClass    obj1(p1=5);
   YellowClass  obj2;
   BlueClass    obj3;
   RedClass     obj4;
equation
   ...
end C2;
```

2.6 EQUATIONS

As we already stated, Modelica is primarily an equation-based language in contrast to ordinary programming languages, where assignment statements proliferate. Equations are more flexible than assignments since they do not prescribe a certain data flow direction or execution order. This is the key to the physical modeling capabilities and increased reuse potential of Modelica classes.

Thinking in equations is a bit unusual for most programmers. In Modelica the following holds:

- Assignment statements in conventional languages are usually represented as equations in Modelica.
- Attribute assignments are represented as equations.
- Connections between objects generate equations.

Equations are more powerful than assignment statements. For example, consider a resistor equation where the resistance R multiplied by the current i is equal to the voltage v:

```
R*i = v;
```

This equation can be used in three ways corresponding to three possible assignment statements: computing the current from the voltage and the resistance, computing the voltage from the resistance and the current, or computing the resistance from the voltage and the current. This is expressed in the following three assignment statements:

```
i := v/R;
v := R*i;
R := v/i;
```

Equations in Modelica can be informally classified into four different groups depending on the syntactic context in which they occur:

- *Normal equations* occurring in equation sections, including the connect equation, which is a special form of equation.
- *Declaration equations*, which are part of variable or constant declarations.
- *Modification equations*, which are commonly used to modify attributes.
- *Initial equations*, specified in initial equation sections or as start attribute equations. These equations are used to solve the initialization problem at startup time.

As we already have seen in several examples, normal equations appear in equation sections started by the keyword equation and terminated by some other allowed keyword:

```
equation
   ...
   <equations>
   ...
<some other allowed keyword>
```

The above resistor equation is an example of a normal equation that can be placed in an equation section. Declaration equations are usually given as part of declarations of fixed or parameter constants, for example:

```
constant Integer one = 1;
parameter Real mass = 22.5;
```

An equation always holds, which means that the mass in the above example never changes value during simulation. It is also possible to specify a declaration equation for a normal variable, for example:

```
Real speed = 72.4;
```

However, this does not make much sense since it will constrain the variable to have the same value throughout the computation, effectively behaving as a constant. Therefore a declaration equation is quite different from a variable initializer in other languages.

Concerning attribute assignments, these are typically specified using modification equations. For example, if we need to specify an initial value for a variable, meaning its value at the start of the computation, then we give an attribute equation for the start attribute of the variable, for example:

```
Real speed(start=72.4);
```

2.6.1 Repetitive Equation Structures

Before reading this section you might want to take a look at Section 2.13 about arrays and Section 2.14.2 about statements and algorithmic for-loops.

Sometimes there is a need to conveniently express sets of equations that have a regular, that is, repetitive structure. Often this can be expressed as array equations, including references to array elements denoted using square bracket notation. However, for the more general case of repetitive equation structures Modelica provides a loop construct. Note that this is not a loop in the algorithmic sense of the word—it is rather a shorthand notation for expressing a set of equations.

For example, consider an equation for a polynomial expression:

```
y = a[1]+a[2]*x + a[3]*x^2 + ... + a[n+1]*x^n
```

The polynomial equation can be expressed as a set of equations with regular structure in Modelica, with y equal to the scalar product of the vectors a and xpowers, both of length n+1:

```
xpowers[1] = 1;
xpowers[2] = xpowers[1]*x;
xpowers[3] = xpowers[2]*x;
...
xpowers[n+1] = xpowers[n]*x;
y = a * xpowers;
```

The regular set of equations involving xpowers can be expressed more conveniently using the for loop notation:

```
for i in 1:n loop
   xpowers[i+1] = xpowers[i]*x;
end for;
```

In this particular case a vector equation provides an even more compact notation:

```
xpowers[2:n+1] = xpowers[1:n]*x;
```

Here the vectors x and xpowers have length n+1. The colon notation 2:n+1 means extracting a vector of length n, starting from element 2 up to and including element n+1.

2.6.2 Partial Differential Equations

Partial differential equations (PDEs) contain derivatives with respect to other variables than time, for example, of spatial Cartesian coordinates such as x and y. Models of phenomena such as heat flow or fluid flow typically involve PDEs. PDE functionality is not yet part of the official Modelica language but is planned for the future.

2.7 ACAUSAL PHYSICAL MODELING

Acausal modeling is a declarative modeling style, meaning modeling based on equations instead of assignment statements. Equations do not specify which variables are inputs and which are outputs, whereas in assignment statements variables on the left-hand side are always outputs (results) and variables on the right-hand side are always inputs. Thus, the causality of equation-based models is unspecified and becomes fixed only when the corresponding equation systems are solved. This is called *acausal modeling*. The term *physical modeling* reflects the fact that *acausal modeling* is very well suited for representing the *physical structure* of modeled systems.

The main advantage with acausal modeling is that the solution direction of equations will adapt to the data flow context in which the solution is computed. The data flow context is defined by stating which variables are needed as *outputs* and which are external *inputs* to the simulated system.

The acausality of Modelica library classes makes these more reusable than traditional classes containing assignment statements where the `input-output` causality is fixed.

2.7.1 Physical Modeling Versus Block-Oriented Modeling

To illustrate the idea of acausal physical modeling we give an example of a simple electrical circuit (Fig. 2.7). The connection diagram[6] of the electrical circuit shows how the components are connected. It may be drawn with component placements to roughly correspond to the physical layout of the electrical circuit on a printed circuit board. The physical connections in the real circuit correspond to the logical connections in the diagram. Therefore the term *physical modeling* is quite appropriate.

The Modelica `SimpleCircuit` model below directly corresponds to the circuit depicted in the connection diagram of Figure 2.7. Each graphic object in the diagram corresponds to a declared instance in the simple circuit model. The model is acausal since no signal flow, that is, cause-and-effect flow, is specified. Connections between objects are specified using the `connect` equation construct, which is a special syntactic form of equation that we will examine later. The classes `Resistor`, `Capacitor`, `Inductor`, `VsourceAC`, and `Ground` will be presented in more detail in Sections 2.11 and 2.12.

```
model SimpleCircuit
  Resistor   R1(R=10);
  Capacitor  C(C=0.01);
```

-----

[6]A connection diagram emphasizes the connections between components of a model, whereas a composition diagram specifies which components a model is composed of, their subcomponents, and so forth. A class diagram usually depicts inheritance and composition relations.

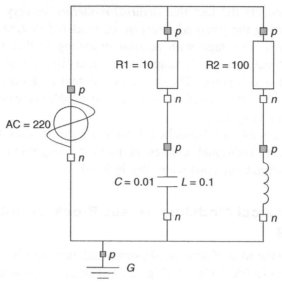

Figure 2.7   Connection diagram of the acausal simple circuit model.

```
    Resistor   R2(R=100);
    Inductor   L(L=0.1);
    VsourceAC AC;
    Ground     G;
equation
    connect(AC.p, R1.p);   // Capacitor circuit
    connect(R1.n, C.p);
    connect(C.n, AC.n);
    connect(R1.p, R2.p);   // Inductor circuit
    connect(R2.n, L.p);
    connect(L.n, C.n);
    connect(AC.n, G.p);    // Ground
end SimpleCircuit;
```

As a comparison we show the same circuit modeled using causal block-oriented modeling depicted as a diagram in Figure 2.8. Here the physical topology is lost—the structure of the diagram has no simple correspondence to the structure of the physical circuit board. This model is causal since the signal flow has been deduced and is clearly shown in the diagram. Even for this simple example the analysis to convert the intuitive physical model to a causal block-oriented model is nontrivial. Another disadvantage is that the resistor representations are context dependent. For example, the resistors R1 and R2 have

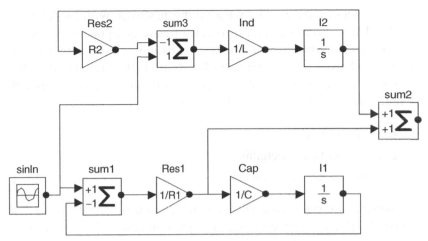

Figure 2.8 Simple circuit model using causal block-oriented modeling with explicit signal flow.

different definitions, which makes reuse of model library components hard. Furthermore, such system models are usually hard to maintain since even small changes in the physical structure may result in large changes to the corresponding block-oriented system model.

2.8 THE MODELICA SOFTWARE COMPONENT MODEL

For a long time, software developers have looked with envy on hardware system builders, regarding the apparent ease with which reusable hardware components are used to construct complicated systems. With software there seems too often to be a need or tendency to develop from scratch instead of reusing components. Early attempts at software components include procedure libraries, which unfortunately have too limited applicability and low flexibility. The advent of object-oriented programming has stimulated the development of software component frameworks such as CORBA, the Microsoft COM/DCOM component object model, and JavaBeans. These component models have considerable success in certain application areas, but there is still a long way to go to reach the level of reuse and component standardization found in hardware industry.

The reader might wonder what all this has to do with Modelica. In fact, Modelica offers quite a powerful software component model that is on par with hardware component systems in flexibility and potential for reuse. The key to this increased flexibility is the fact that Modelica classes are based on equations. What is a software component model? It should include the following three items:

1. Components
2. A connection mechanism
3. A component framework

Components are connected via the connection mechanism, which can be visualized in connection diagrams. The component framework realizes components and connections and ensures that communication works and constraints are maintained over the connections. For systems composed of acausal components the direction of data flow, that is, the causality is automatically deduced by the compiler at composition time.

2.8.1 Components

Components are simply instances of Modelica classes. Those classes should have well-defined interfaces, sometimes called ports, in Modelica called connectors, for communication and coupling between a component and the outside world.

A component is modeled independently of the environment where it is used, which is essential for its reusability. This means that in the definition of the component including its equations, only local variables and connector variables can be used. No means of communication between a component and the rest of the system, apart from going via a connector, should be allowed. However, in Modelica access of component data via dot notation is also possible. A component may internally consist of other connected components, that is, hierarchical modeling.

2.8.2 Connection Diagrams

Complex systems usually consist of large numbers of connected components, of which many components can be hierarchically decomposed into other components through several levels. To grasp this complexity, a pictorial representation of components and connections is quite important. Such graphic representation is available as connection diagrams, of which a schematic example is shown in Figure 2.9. We have earlier presented a connection diagram of a simple circuit in Figure 2.7.

Each rectangle in the diagram example represents a physical component, for example, a resistor, a capacitor, a transistor, a mechanical gear, a valve, and so forth. The connections represented by lines in the diagram correspond to real, physical connections. For example, connections can be realized by electrical wires, by the mechanical connections, by pipes for fluids, by heat exchange between components, and the like. The connectors, that is, interface points, are shown as small square dots on the rectangle in the diagram. Variables at such interface points define the interaction between the component represented by the rectangle and other components. A simple car example of a connection diagram for an application in the mechanical domain is shown in Figure 2.10.

The simple car model below includes variables for subcomponents such as wheels, chassis, and control unit. A "comment" string after

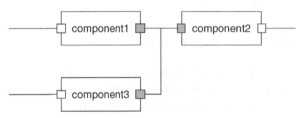

Figure 2.9   Schematic picture of connection diagram for components.

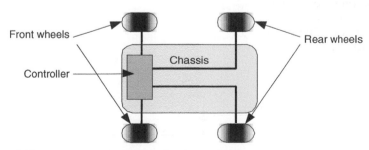

Figure 2.10   Connection diagram for simple car model.

the class name briefly describes the class. The wheels are connected to both the chassis and the controller. Connect equations are present but are not shown in this partial example.

```
class Car   "A car class to combine car components"
   Wheel          w1,w2,w3,w4   "Wheel one to four";
   Chassis        chassis       "Chassis";
   CarController  controller    "Car controller";
   ...
end Car;
```

2.8.3 Connectors and Connector Classes

Modelica connectors are instances of connector classes, which define the variables that are part of the communication interface that is specified by a connector. Thus, connectors specify external interfaces for interaction.

For example, Pin is a connector class that can be used to specify the external interfaces for electrical components (Fig. 2.11) that have pins. The types Voltage and Current used within Pin are the same as Real but with different associated units. From the Modelica language point of view the types Voltage and Current are similar to Real and are regarded as having equivalent types. Checking unit compatibility within equations is optional.

```
type Voltage = Real(unit="V");
type Current = Real(unit="A");
```

The Pin connector class below contains two variables. The flow prefix on the second variable indicates that this variable represents a flow quantity, which has special significance for connections as explained in the next section.

```
connector Pin
   Voltage        v;
   flow Current   i;
end Pin;
```

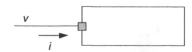

Figure 2.11   Component with one electrical Pin connector.

2.8.4 Connections

Connections between components can be established between connectors of equivalent type. Modelica supports equation-based acausal connections, which means that connections are realized as equations. For acausal connections, the direction of data flow in the connection need not be known. Additionally, causal connections can be established by connecting a connector with an `output` attribute to a connector declared as `input`.

Two types of coupling can be established by connections, depending on whether the variables in the connected connectors are nonflow (default) or declared using the `flow` prefix:

1. Equality coupling, for nonflow variables, according to Kirchhoff's first law
2. Sum-to-zero coupling, for flow variables, according to Kirchhoff's current law

For example, the keyword `flow` for the variable `i` of type `Current` in the `Pin` connector class indicates that all currents in connected pins are summed to zero, according to Kirchhoff's current law.

Connection equations are used to connect instances of connection classes. A connection equation `connect(R1.p,R2.p)`, with `R1.p` and `R2.p` of connector class `Pin`, connects the two pins (Fig. 2.12) so that they form one node. This produces two equations, namely:

```
R1.p.v = R2.p.v
R1.p.i + R2.p.i = 0
```

The first equation says that the voltages of the connected wire ends are the same. The second equation corresponds to Kirchhoff's second law, saying that the currents sum to zero at a node (assuming positive value while flowing into the component). The sum-to-zero equations are generated when the prefix `flow` is used. Similar laws apply to

Figure 2.12   Connecting two components that have electrical pins.

flows in piping networks and to forces and torques in mechanical systems.

2.8.5 Implicit Connections with Inner/Outer

So far we have focused on explicit connections between connectors where each connection is explicitly represented by a connect equation and a corresponding line in a connection diagram. However, when modeling certain kinds of large models with many interacting components, this approach becomes rather clumsy because of the large number of potential connections—a connection might be needed between each pair of components. This is especially true for system models involving *force fields*, which lead to a maximum of $n \times n$ connections between the n components influenced by the force field or $1 \times n$ connections between a central object and n components if intercomponent interaction is neglected.

For the case of $1 \times n$ connections, instead of using a large number of explicit connections, Modelica provides a convenient mechanism for *implicit connections* between an object and n of its components through the inner and outer declaration prefixes.

A rather common kind of implicit interaction is where a *shared attribute* of a single environment object is *accessed* by a number of components within that environment. For example, we might have an environment including house components, each accessing a shared environment temperature, or a circuit board environment with electronic components accessing the board temperature.

A Modelica environment component example model along these lines is shown below, where a shared environment temperature variable T0 is declared as a *definition declaration* marked by the keyword inner. This declaration is implicitly accessed by the *reference declarations* of T0 marked by the prefix outer in the components comp1 and comp2.

```
model Environment
  import Modelica.Math.sin;
  inner Real T0;
    //Definition of actual environment temperature T0
  Component comp1, comp2;
    //Lookup match comp1.T0 = comp2.T0 = T0
  parameter Real k=1;
equation
  T0 = sin(k*time);
end Environment;
```

```
model Component
  outer Real T0;
    // Reference to temperature T0 defined in the environments
  Real T;
equation
  T = T0;
end Component;
```

2.8.6 Expandable Connectors for Information Buses

It is common in engineering to have so-called information buses with the purpose to transport information between various system components, for example, sensors, actuators, and control units. Some buses are even standardized (e.g., by IEEE), but usually rather generic to allow many kinds of different components.

This is the key idea behind the expandable connector construct in Modelica. An expandable connector acts like an information bus since it is intended for connection to many kinds of components. To make this possible it automatically expands the expandable connector type to accommodate all the components connected to it with their different interfaces. If an element with a certain name and type is not present, it is added.

All fields in an expandable connector are seen as connector instances even if they are not declared as such, that is, it is possible to connect to for example, a Real variable.

Moreover, when two expandable connectors are connected, each is augmented with the variables that are only declared in the other expandable connector. This is repeated until all connected expandable connector instances have matching variables, that is, each of the connector instances is expanded to be the union of all connector variables. If a variable appears as an input in one expandable connector, it should appear as a noninput in at least one other expandable connector instance in the connected set. The following is a small example:

```
expandable connector EngineBus
end EngineBus;

block Sensor
  RealOutput speed;
end Sensor;

block Actuator
```

```
    RealInput speed;
end Actuator;

model Engine
  EngineBus bus;
  Sensor sensor;
  Actuator actuator;
equation
  connect(bus.speed, sensor.speed);
    // provides the non-input
  connect(bus.speed, actuator.speed);
end Engine;
```

There are many more issues to consider when using expandable connectors; see, for example, Modelica (2010) and Fritzson (2011).

2.8.7 Stream Connectors

In thermodynamics with fluid applications where there can be bi-directional flows of matter with associated quantities, it turns out that the two basic variable types in a connector—potential/nonflow variables and flow variables—are not sufficient to describe models that result in a numerically sound solution approach. Such applications typically have bi-directional flow of matter with convective transport of specific quantities, such as specific enthalpy and chemical composition.

If we would use conventional connectors with flow and nonflow variables, the corresponding models would include nonlinear systems of equations with Boolean unknowns for the flow directions and singularities around zero flow. Such equation systems cannot be solved reliably in general. The model formulations can be simplified when formulating two different balance equations for the two possible flow directions. This is, however, not possible only using flow and nonflow variables.

This fundamental problem is addressed in Modelica by introducing a third type of connector variable, called *stream variable*, declared with the prefix stream. A stream variable describes a quantity that is carried by a flow variable, that is, a purely convective transport phenomenon.

If at least one variable in a connector has the `stream` prefix, the connector is called a *stream connector* and the corresponding variable is called a *stream variable*. For example:

```
connector FluidPort
  ...
  flow  Real m_flow
    "Flow of matter; m_flow>0 if flow into component";
  stream Real h_outflow
    "Specific variable in component if m_flow < 0"
end FluidPort

model FluidSystem
  ...
  FluidComponent m₁, m₂, ..., mₙ;
  FluidPort      c₁, c₂, ..., cₘ;
equation
  connect(m₁.c, m₂.c);
  ...
  connect(mᵢ.c, cₘ);
  ...
end FluidSystem;
```

For more details and further explanations, see Modelica (2010) and Fritzson (2011).

2.9 PARTIAL CLASSES

A common property of many electrical components is that they have two pins. This means that it is useful to define a "blueprint" model class, for example, called `TwoPin`, that captures this common property. This is a *partial class* since it does not contain enough equations to completely specify its physical behavior and is therefore prefixed by the keyword `partial`. Partial classes are usually known as *abstract classes* in other object-oriented languages.

```
partial class TwoPin[7]
  "Superclass of elements with two electrical pins"
  Pin      p, n;
  Voltage  v;
  Current  i;
equation
  v = p.v - n.v;
```

---

[7]This `TwoPin` class is referred to by the name `Modelica.Electrical.Analog.Interfaces.OnePort` in the Modelica standard library since this is the name used by electrical modeling experts. Here we use the more intuitive name `TwoPin` since the class is used for components with two physical ports and not one. The `OnePort` naming is more understandable if it is viewed as denoting composite ports containing two subports.

```
  0 = p.i + n.i;
  i = p.i;
end TwoPin;
```

The TwoPin class has two pins, p and n, a quantity v that defines the voltage drop across the component, and a quantity i that defines the current into pin p, through the component, and out from pin n (Fig. 2.13). It is useful to label the pins differently, for example, p and n, and using graphics, for example, filled and unfilled squares, respectively, to obtain a well-defined sign for v and i, although there is no physical difference between these pins in reality.

The equations define generic relations between quantities of simple electrical components. In order to be useful, a constitutive equation must be added that describes the specific physical characteristics of the component.

2.9.1 Reuse of Partial Classes

Given the generic partial class TwoPin, it is now straightforward to create the more specialized Resistor class by adding a constitutive equation:

```
  R*i = v;
```

This equation describes the specific physical characteristics of the relation between voltage and current for a resistor (Fig. 2.14).

```
class Resistor "Ideal electrical resistor"
  extends TwoPin;
  parameter Real R(unit="Ohm") "Resistance";
equation
```

Figure 2.13 Generic TwoPin class that describes the general structure of simple electrical components with two pins.

Figure 2.14 Resistor component.

```
   R*i = v;
 end Resistor;
```

A class for electrical capacitors can also reuse TwoPin in a similar way, adding the constitutive equation for a capacitor (Fig. 2.15).

```
class Capacitor "Ideal electrical capacitor"
  extends TwoPin;
  parameter Real C(Unit="F") "Capacitance";
equation
  C*der(v) = i;
end Capacitor;
```

During system simulation the variables i and v specified in the above components evolve as functions of time. The solver of differential equations computes the values of $v(t)$ and $i(t)$ (where t is time) such that $C \cdot \dot{v}(t) = i(t)$ for all values of t, fulfilling the constitutive equation for the capacitor.

2.10 COMPONENT LIBRARY DESIGN AND USE

In a similar way as we previously created the resistor and capacitor components, additional electrical component classes can be created, forming a simple electrical component library that can be used for application models such as the SimpleCircuit model. Component libraries of reusable components are actually the key to effective modeling of complex systems.

2.11 EXAMPLE: ELECTRICAL COMPONENT LIBRARY

Below we show an example of designing a small library of electrical components needed for the simple circuit example, as well as the equations that can be extracted from these components.

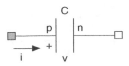

Figure 2.15 Capacitor component.

2.11.1 Resistor

Four equations can be extracted from the resistor model depicted in Figures 2.14 and 2.16. The first three originate from the inherited TwoPin class, whereas the last is the constitutive equation of the resistor.

```
0 = p.i + n.i
v = p.v - n.v
i = p.i
v = R*i
```

2.11.2 Capacitor

The following four equations originate from the capacitor model depicted in Figures 2.15 and 2.17, where the last equation is the constitutive equation for the capacitor.

```
0 = p.i + n.i
v = p.v - n.v
i = p.i
i = C * der(v)
```

2.11.3 Inductor

The inductor class depicted in Figure 2.18 and shown below gives a model for ideal electrical inductors.

```
class Inductor "Ideal electrical inductor"
   extends TwoPin;
```

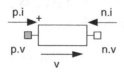

Figure 2.16 Resistor component.

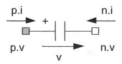

Figure 2.17 Capacitor component.

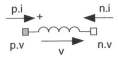

Figure 2.18   Inductor component.

```
   parameter Real L(unit="H") "Inductance";
equation
   v = L*der(i);
end Inductor;
```

These equations can be extracted from the inductor class, where the first three come from TwoPin as usual and the last is the constitutive equation for the inductor.

```
0 = p.i + n.i
v = p.v - n.v
i = p.i
v = L * der(i)
```

2.11.4 Voltage Source

A class VsourceAC for the sine-wave voltage source to be used in our circuit example is depicted in Figure 2.19 and can be defined as below. This model as well as other Modelica models specify behavior that evolves as a function of time. Note that a predefined variable time is used. In order to keep the example simple, the constant PI is explicitly declared even though it is usually imported from the Modelica standard library.

```
class VsourceAC "Sin-wave voltage source"
  extends TwoPin;
  parameter Voltage  VA         = 220 "Amplitude";
  parameter Real     f(unit="Hz") = 50  "Frequency";
  constant  Real     PI         = 3.141592653589793;
equation
```

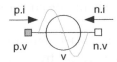

Figure 2.19   Voltage source component VsourceAC, where v(t) = VA*sin(2*PI*f*time).

Figure 2.20 Ground component.

```
  v = VA*sin(2*PI*f*time);
end VsourceAC;
```

In this `TwoPin`-based model, four equations can be extracted from the model, of which the first three are inherited from `TwoPin`:

```
0 = p.i + n.i
v = p.v - n.v
i = p.i
v = VA*sin(2*PI*f*time)
```

2.11.5 Ground

Finally, we define a class for ground points that can be instantiated as a reference value for the voltage levels in electrical circuits. This class has only one pin (Fig. 2.20).

```
class Ground "Ground"
  Pin p;
equation
  p.v = 0;
end Ground;
```

A single equation can be extracted from the `Ground` class.

```
p.v = 0
```

2.12 SIMPLE CIRCUIT MODEL

Having collected a small library of simple electrical components, we can now put together the simple electrical circuit shown previously and in Figure 2.21.

The two resistor instances `R1` and `R2` are declared with modification equations for their respective resistance parameter values. Similarly, an instance `C` of the capacitor and an instance `L` of the inductor are declared with modifiers for capacitance and inductance,

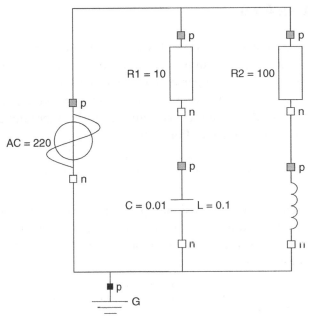

Figure 2.21   Simple circuit model.

respectively. The voltage source AC and the ground instance G have no modifiers. Connect equations are provided to connect the components in the circuit.

```
class SimpleCircuit
   Resistor   R1(R=10);
   Capacitor  C(C=0.01);
   Resistor   R2(R=100);
   Inductor   L(L=0.1);
   VsourceAC  AC;
   Ground     G;
equation
   connect(AC.p, R1.p);   // 1, Capacitor circuit
   connect(R1.n, C.p);    //      Wire 2
   connect(C.n, AC.n);    //      Wire 3
   connect(R1.p, R2.p);   // 2, Inductor circuit
   connect(R2.n, L.p);    //      Wire 5
   connect(L.n, C.n);     //      Wire 6
   connect(AC.n, G.p);    // 7, Ground
end SimpleCircuit;
```

2.13 ARRAYS

An array is a collection of variables all of the same type. Elements of
an array are accessed through simple integer indexes, ranging from a
lower bound of 1 to an upper bound being the size of the respective
dimension. An array variable can be declared by appending dimensions
within square brackets after a class name, as in Java, or after a variable
name, as in the C language. For example:

```
Real[3]      positionvector = {1,2,3};
Real[3,3]    identitymatrix = {{1,0,0}, {0,1,0}, {0,0,1}};
Real[3,3,3] arr3d;
```

This declares a three-dimensional position vector, a transformation
matrix, and a three-dimensional array. Using the alternative syntax of
attaching dimensions after the variable name, the same declarations
can be expressed as:

```
Real  positionvector[3]   = {1,2,3};
Real  identitymatrix[3,3] = {{1,0,0}, {0,1,0}, {0,0,1}};
Real  arr3d[3,3,3];
```

In the first two array declarations, declaration equations have been
given, where the array constructor `{}` is used to construct array values
for defining `positionvector` and `identitymatrix`. Indexing of
an array A is written A[i,j,...], where 1 is the lower bound and
size(A,k) is the upper bound of the index for the kth dimension.
Submatrices can be formed by utilizing the : notation for index ranges,
for example, A[i1:i2, j1:j2], where a range i1:i2 means all
indexed elements starting with i1 up to and including i2.

Array expressions can be formed using the arithmetic operators
+, −, *, and /, since these can operate on either scalars, vectors,
matrices, or (when applicable) multidimensional arrays with elements
of type Real or Integer. The multiplication operator * denotes
scalar product when used between vectors, matrix multiplication when
used between matrices or between a matrix and a vector, and elementwise multiplication when used between an array and a scalar.
As an example, multiplying `positionvector` by the scalar 2 is
expressed by

```
positionvector*2
```

which gives the result

`{2,4,6}`

In contrast to Java, arrays of dimensionality >1 in Modelica are always rectangular as in Matlab or Fortran.

A number of built-in array functions are available, of which a few are shown in the following list:

`transpose(A)`	Permutes the first two dimensions of array A.
`zeros(n1,n2,n3,...)`	Returns an $n_1 \times n_2 \times n_3 \times \ldots$ zero-filled integer array.
`ones(n1,n2,n3,...)`	Returns an $n_1 \times n_2 \times n_3 \times \ldots$ one-filled integer array.
`fill(s,n1,n2,n3,...)`	Returns the $n_1 \times n_2 \times n_3 \times \ldots$ array with all elements filled with the value of the scalar expression s.
`min(A)`	Returns the smallest element of array expression A.
`max(A)`	Returns the largest element of array expression A.
`sum(A)`	Returns the sum of all the elements of array expression A.

A scalar Modelica function of a scalar argument is automatically generalized to be applicable also to arrays elementwise. For example, if A is a vector of real numbers, then `cos(A)` is a vector where each element is the result of applying the function `cos` to the corresponding element in A. For example:

`cos({1, 2, 3}) = {cos(1), cos(2), cos(3)}`

General array concatenation can be done through the array concatenation operator `cat(k,A,B,C,...)` that concatenates the arrays `A,B,C,...` along the k:th dimension. For example, `cat(1,{2,3}, {5,8,4})` gives the result `{2,3,5,8,4}`.

The common special cases of concatenation along the first and second dimensions are supported through the special syntax forms `[A;B;C;...]` and `[A,B,C,...]`, respectively. Both of these forms can be mixed. In order to achieve compatibility with Matlab array syntax, being a de facto standard, scalar and vector arguments to these special operators are promoted to become matrices before performing the concatenation. This gives the effect that a matrix can be constructed from scalar expressions by separating rows by semicolons and columns by commas. The example below creates an $m \times n$ matrix:

```
[expr₁₁, expr₁₂, ... expr₁ₙ;
 expr₂₁, expr₂₂, ... expr₂ₙ;
 ...
 exprₘ₁, exprₘ₂, ... exprₘₙ]
```

It is instructive to follow the process of creating a matrix from scalar expressions using these operators. For example:

```
[1,2;
 3,4]
```

First, each scalar argument is promoted to become a matrix, giving

```
[{{1}}, {{2}};
 {{3}}, {{4}}]
```

Since [... , ...] for concatenation along the second dimension has higher priority than [... ; ...], which concatenates along the first dimension, the first concatenation step gives

```
[{{1, 2}};
 {{3, 4}}]
```

Finally, the row matrices are concatenated giving the desired 2×2 matrix:

```
{{1, 2}},
 {3, 4}}
```

The special case of just one scalar argument can be used to create a 1×1 matrix. For example:

```
[1]
```

gives the matrix

```
{{1}}
```

2.14 ALGORITHMIC CONSTRUCTS

Even though equations are eminently suitable for modeling physical systems and for a number of other tasks, there are situations where nondeclarative algorithmic constructs are needed. This is typically the case for algorithms, that is, procedural descriptions of how to carry out specific computations, usually consisting of a number of statements that should be executed in the specified order.

2.14.1 Algorithm Sections and Assignment Statements

In Modelica, algorithmic statements can occur only within algorithm sections, starting with the keyword `algorithm`. Algorithm sections may also be called algorithm equations, since an algorithm section can be viewed as a group of equations involving one or more variables, and can appear among equation sections. Algorithm sections are terminated by the appearance of one of the keywords `equation`, `public`, `protected`, `algorithm`, or `end`.

```
algorithm
   ...
   <statements>
   ...
   <some other keyword>
```

An algorithm section embedded among equation sections can appear as below, where the example algorithm section contains three assignment statements.

```
equation
   x = y*2;
   z = w;
algorithm
   x1 := z+x;
   x2 := y-5;
   x1 := x2+y;
equation
   u = x1+x2;
   ...
```

Note that the code in the algorithm section, sometimes denoted algorithm equation, uses the values of certain variables from outside the algorithm. These variables are so-called *input variables* to the algorithm—in this example x, y, and z. Analogously, variables assigned values by the algorithm define the *outputs of the algorithm* — in this example x1 and x2. This makes the semantics of an algorithm section quite similar to a function with the algorithm section as its body, and with input and output formal parameters corresponding to inputs and outputs as described above.

2.14.2 Statements

In addition to assignment statements, which were used in the previous example, a few other kinds of "algorithmic" statements are available in Modelica: if-then-else statements, for loops, while loops, return statements, and so on. The summation below uses both a while loop and an if statement, where size(a,1) returns the size of the first dimension of array a. The elseif and else parts of if statements are optional.

```
sum := 0;
n := size(a,1);
while n>0 loop
  if a[n]>0 then
    sum := sum + a[n];
  elseif a[n] > -1 then
    sum := sum - a[n] -1;
  else
    sum := sum - a[n];
  end if;
  n := n-1;
end while;
```

Both for loops and while loops can be immediately terminated by executing a break statement inside the loop. Such a statement just consists of the keyword break followed by a semicolon.

Consider once more the computation of the polynomial presented in Section 2.6.1 on repetitive equation structures.

```
y := a[1]+a[2]*x + a[3]*x^1 + ... + a[n+1]*x^n;
```

When using equations to model the computation of the polynomial, it was necessary to introduce an auxiliary vector xpowers for storing the different powers of x. Alternatively, the same computation can be expressed as an algorithm including a for loop as below. This can be done without the need for an extra vector — it is enough to use a scalar variable xpower for the most recently computed power of x.

```
algorithm
  y := 0;
  xpower := 1;
  for i in 1:n+1 loop
    y := y + a[i]*xpower;
    xpower := xpower*x;
  end for;
  ...
```

2.14.3 Functions

Functions are a natural part of any mathematical model. A number of mathematical functions like `abs`, `sqrt`, `mod`, and the like are predefined in the Modelica language, whereas others such as `sin`, `cos`, `exp`, and the like are available in the Modelica standard math library `Modelica.Math`. The arithmetic operators $+, -, *, /$ can be regarded as functions that are used through a convenient operator syntax. Thus, it is natural to have user-defined mathematical functions in the Modelica language. The body of a Modelica function is an algorithm section that contains procedural algorithmic code to be executed when the function is called. Formal parameters are specified using the `input` keyword, whereas results are denoted using the `output` keyword. This makes the syntax of function definitions quite close to Modelica block class definitions.

Modelica functions are *mathematical functions*, that is, without global side effects and with no memory. A Modelica function always returns the same results given the same arguments. Below we show the algorithmic code for polynomial evaluation in a function named `polynomialEvaluator`.

```
function polynomialEvaluator
  input  Real a[:];
    // Array, size defined at function call time
  input  Real x := 1.0;
    // Default value 1.0 for x
  output Real y;
protected
  Real   xpower;
algorithm
  y := 0;
  xpower := 1;
  for i in 1:size(a,1) loop
```

```
    y := y + a[i]*xpower;
    xpower := xpower*x;
  end for;
end polynomialEvaluator;
```

Functions are usually called with positional association of actual arguments to formal parameters. For example, in the call below the actual argument {1,2,3,4} becomes the value of the coefficient vector a, and 21 becomes the value of the formal parameter x. Modelica function parameters are read-only, that is, they may not be assigned values within the code of the function. When a function is called using positional argument association, the number of actual arguments and formal parameters must be the same. The types of the actual argument expressions must be compatible with the declared types of the corresponding formal parameters. This allows passing array arguments of arbitrary length to functions with array formal parameters with unspecified length, as in the case of the input formal parameter a in the polynomialEvaluator function.

```
  p = polynomialEvaluator({1, 2, 3, 4}, 21);
```

The same call to the function polynomialEvaluator can instead be made using named association of actual arguments to formal parameters, as in the next example. This has the advantage that the code becomes more self-documenting as well as more flexible with respect to code updates.

For example, if all calls to the function polynomialEvaluator are made using named parameter association, the order between the formal parameters a and x can be changed, and new formal parameters with default values can be introduced in the function definitions without causing any compilation errors at the call sites. Formal parameters with default values need not be specified as actual arguments unless those parameters should be assigned values different from the defaults.

```
  p = polynomialEvaluator(a={1, 2, 3, 4}, x=21);
```

Functions can have multiple results. For example, the function f below has three result parameters declared as three formal output parameters r1, r2, and r3.

```
  function f
    input Real x;
```

```
  input Real y;
  output Real r1;
  output Real r2;
  output Real r3;
  ...
end f;
```

Within algorithmic code multiresult functions may be called only in special assignment statements, as the one below, where the variables on the left-hand side are assigned the corresponding function results.

```
(a, b, c) := f(1.0, 2.0);
```

In equations a similar syntax is used:

```
(a, b, c) = f(1.0, 2.0);
```

A function is returned by reaching the end of the function or by executing a return statement inside the function body.

2.14.4 Operator Overloading and Complex Numbers

Function and operator overloading allow several definitions of the same function or operator, but with a different set of input formal parameter types for each definition. This allows, for example, to define operators such as addition, multiplication, and so forth, of complex numbers, using the ordinary + and * operators but with new definitions, or provide several definitions of a `solve` function for linear matrix equation solution for different matrix representations such as standard dense matrices, sparse matrices, symmetric matrices, and so forth.

In fact, overloading already exists predefined to a limited extent for certain operators in the Modelica language. For example, the plus (+) operator for addition has several different definitions depending on the data type:

- 1+2 means integer addition of two integer constants giving an integer result, here 3.
- 1.0+2.0 means floating-point number addition of two `Real` constants giving a floating-point number result, here 3.0.

- "ab"+"2" means string concatenation of two string constants giving a string result, here "ab2".
- {1,2}+{3,4} means integer vector addition of two integer constant vectors giving a vector result, here {4,6}.

Overloaded operators for user-defined data types can be defined using `operator record` and `operator function` declarations. Here we show part of a complex numbers data type example:

```
operator record Complex "Record defining a Complex number"

  Real re "Real part of complex number";
  Real im "Imaginary part of complex number";

  encapsulated operator 'constructor'
   import Complex;

    function fromReal
      input  Real re;
      output Complex result = Complex(re=re, im=0.0);
        annotation(Inline=true);
    end fromReal;
  end 'constructor';

  encapsulated operator function '+'
   import Complex;
    input  Complex c1;
    input  Complex c2;
    output Complex result  "Same as: c1 + c2";
      annotation(Inline=true);
  algorithm
    result := Complex(c1.re + c2.re, c1.im + c2.im);
  end '+';

end Complex;
```

In the above example we start as usual with the real and imaginary part declarations of the re and im fields of the Complex operator record definition. Then comes a *constructor definition* fromReal with only one input argument instead of the two inputs of the default Complex constructor implicitly defined by the Complex record definition, followed by overloaded operator definition for '+'.

How can these definitions be used? Take a look at the following small example:

```
Real    a;
Complex b;
Complex c = a + b;
  // Addition of Real number a and Complex number b
```

The interesting part is in the third line, which contains an addition a+b of a Real number a and a Complex number b. There is no built-in addition operator for complex numbers, but we have the above overloaded operator definition of '+' for two complex numbers. An addition of two complex numbers would match this definition right away in the lookup process.

However, in this case we have an addition of a real number and a complex number. Fortunately, the lookup process for overloaded binary operators can also handle this case if there is a constructor function in the Complex record definition that can convert a real number to a complex number. Here we have such a constructor called fromReal.

Note that Complex is predefined in a Modelica library so that it can be used directly.

2.14.5 External Functions

It is possible to call functions defined outside of the Modelica language, implemented in C or Fortran. If no external language is specified, the implementation language is assumed to be C. The body of an external function is marked with the keyword external in the Modelica external function declaration.

```
function log
   input Real x;
   output Real y;
external
end log;
```

The external function interface supports a number of advanced features such as in–out parameters, local work arrays, external function argument order, explicit specification of row-major versus column-major array memory layout, and the like. For example, the formal parameter Ares corresponds to an in–out parameter in the external function leastSquares below, which has the value A as input default and a different value as the result. It is possible to control the ordering and usage of parameters to the function external to Modelica. This is used below to explicitly pass sizes of array dimensions to the Fortran routine called dgels. Some old-style Fortran routines like dgels need work arrays, which is conveniently handled by local variable declarations after the keyword protected.

```
function leastSquares "Solves a linear least squares problem"
   input  Real A[:,:];
```

```
   input   Real B[:,:];
   output Real Ares[size(A,1),size(A,2)] := A;
      //Factorization is returned in Ares for later use
   output Real x[size(A,2),size(B,2)];
protected
   Integer lwork = min(size(A,1),size(A,2))+
                    max(max(size(A,1),size(A,2)),size(B,2))*32;
   Real work[lwork];
   Integer info;
   String transposed="NNNN";
      // Workaround for passing CHARACTER data to
      // Fortran routine
   external "FORTRAN 77"
   dgels(transposed, 100, size(A,1), size(A,2), size(B,2), Ares,
         size(A,1), B, size(B,1), work, lwork, info);
end leastSquares;
```

2.14.6 Algorithms Viewed as Functions

The function concept is a basic building block when defining the semantics or meaning of programming language constructs. Some programming languages are completely defined in terms of mathematical functions. This makes it useful to try to understand and define the semantics of algorithm sections in Modelica in terms of functions. For example, consider the algorithm section below, which occurs in an equation context:

```
algorithm
   y := x;
   z := 2*y;
   y := z+y;
   ...
```

This algorithm can be transformed into an equation and a function as below, without changing its meaning. The equation equates the output variables of the previous algorithm section with the results of the function f. The function f has the inputs to the algorithm section as its input formal parameters and the outputs as its result parameters. The algorithmic code of the algorithm section has become the body of the function f.

```
(y,z) = f(x);
...
function f
   input  Real x;
   output Real y,z;
algorithm
```

```
    y := x;
    z := 2*y;
    y := z+y;
end f;
```

2.15 DISCRETE EVENT AND HYBRID MODELING

Macroscopic physical systems in general evolve continuously as a function of time, obeying the laws of physics. This includes the movements of parts in mechanical systems, current and voltage levels in electrical systems, chemical reactions, and so forth. Such systems are said to have continuous dynamics.

On the other hand, it is sometimes beneficial to make the approximation that certain system components display discrete behavior, that is, changes of values of system variables may occur instantaneously and discontinuously at specific points in time.

In the real physical system the change can be very fast but not instantaneous. Examples are collisions in mechanical systems, for example, a bouncing ball that almost instantaneously changes direction, switches in electrical circuits with quickly changing voltage levels, valves and pumps in chemical plants, and the like. We talk about system components with discrete-time dynamics. The reason to make the discrete approximation is to simplify the mathematical model of the system, making the model more tractable and usually speeding up the simulation of the model several orders of magnitude.

For this reason it is possible to have variables in Modelica models of *discrete-time variability*, that is, the variables change value only at specific points in time, so-called *events*, and keep their values constant between events, as depicted in Figure 2.22. Examples of discrete-time variables are `Real` variables declared with the prefix `discrete` or `Integer`, `Boolean`, and `enumeration` variables, which are discrete time by default and cannot be continuous time.

Since the discrete-time approximation can only be applied to certain subsystems, we often arrive at system models consisting of interacting continuous and discrete components. Such a system is called a *hybrid system* and the associated modeling techniques *hybrid modeling*. The introduction of hybrid mathematical models creates new

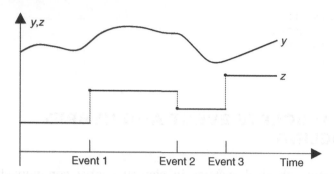

Figure 2.22 Discrete-time variable z changes value only at event instants, whereas continuous-time variables like y may change value both between and at events.

difficulties for their solution, but the disadvantages are far outweighed by the advantages.

Modelica provides two kinds of constructs for expressing hybrid models: conditional expressions or equations to describe discontinuous and conditional models and when equations to express equations that are valid only at discontinuities, for example, when certain conditions become true. For example, if-then-else conditional expressions allow modeling of phenomena with different expressions in different operating regions, as for the equation describing a limiter below.

```
y = if v > limit then limit else v;
```

A more complete example of a conditional model is the model of an ideal diode. The characteristic of a real physical diode is depicted in Figure 2.23, and the ideal diode characteristic in parameterized form is shown in Figure 2.24.

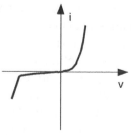

Figure 2.23 Real diode characteristic.

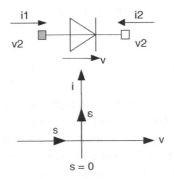

Figure 2.24 Ideal diode characteristic.

Since the voltage level of the ideal diode would go to infinity in an ordinary voltage–current diagram, a parameterized description is more appropriate, where both the voltage v and the current i, same as i1, are functions of the parameter s. When the diode is off, no current flows and the voltage is negative, whereas when it is on, there is no voltage drop over the diode and the current flows.

```
model Diode "Ideal diode"
  extends TwoPin;
  Real s;
  Boolean off;
equation
  off = s < 0;
  if off
    then v=s;
    else v=0;    // conditional equations
  end if;
  i = if off then 0 else s;
    // conditional expression
end Diode;
```

When equations have been introduced in Modelica to express *instantaneous equations*, that is, equations that are valid only at certain points in time that, for example, occur at discontinuities when specific conditions become true, so-called *events*. The syntax of when equations for the case of a vector of conditions is shown below. The equations in the when equation are activated when at least one of the conditions becomes true and remain activated only for a time instant of zero duration. A single condition is also possible.

```
when {condition1, condition2, ...} then
  <equations>
end when;
```

A bouncing ball is a good example of a hybrid system for which the when equation is appropriate when modeled. The motion of the ball is characterized by the variable `height` above the ground and the vertical `velocity`. The ball moves continuously between bounces, whereas discrete changes occur at bounce times, as depicted in Figure 2.25. When the ball bounces against the ground, its velocity is reversed. An ideal ball would have an elasticity coefficient of 1 and would not lose any energy at a bounce. A more realistic ball, as the one modeled below, has an elasticity coefficient of 0.9, making it keep 90% of its speed after the bounce.

The bouncing ball model contains the two basic equations of motion relating height and velocity as well as the acceleration caused by the gravitational force. At the bounce instant the velocity is suddenly reversed and slightly decreased, that is, `velocity` (after bounce) = `-c*velocity` (before bounce), which is accomplished by the special `reinit` syntactic form of instantaneous equation for reinitialization: `reinit (velocity,-c*pre(velocity))`, which in this case reinitializes the `velocity` variable.

```
model BouncingBall "Simple model of a bouncing ball"
  constant  Real g = 9.81    "Gravity constant";
  parameter Real c = 0.9     "Coefficient of restitution";
  parameter Real radius=0.1 "Radius of the ball";
  Real height(start = 1)    "Height of the ball center";
  Real velocity(start = 0)  "Velocity of the ball";
equation
  der(height) = velocity;
  der(velocity) = -g;
  when height <= radius then
```

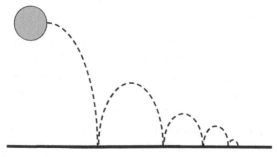

Figure 2.25 Bouncing ball.

```
    reinit(velocity,-c*pre(velocity));
  end when;
end BouncingBall;
```

Note that the equations within a when equation are active only during the instant in time when the condition(s) of the when equation become true, whereas the conditional equations within an if equation are active as long as the condition of the if equation is true.

If we simulate this model long enough, the ball will fall through the ground. This strange behavior of the simulation, called shattering or the Zeno effect, is due to the limited precision of floating-point numbers together with the event detection mechanism of the simulator and occurs for some (unphysical) models where events may occur infinitely close to each other. The real problem in this case is that the model of the impact is not realistic—the law new_velocity = -c*velocity does not hold for very small velocities. A simple fix is to state a condition when the ball falls through the ground and then switch to an equation stating that the ball is lying on the ground. A better but more complicated solution is to switch to a more realistic material model.

2.16 PACKAGES

Name conflicts are a major problem when developing reusable code, for example, libraries of reusable Modelica classes and functions for various application domains. No matter how carefully names are chosen for classes and variables it is likely that someone else is using some name for a different purpose. This problem gets worse if we are using short descriptive names since such names are easy to use and therefore quite popular, making them quite likely to be used in another person's code.

A common solution to avoid name collisions is to attach a short prefix to a set of related names, which are grouped into a package. For example, all names in the X-Windows toolkit have the prefix Xt, and WIN32 is the prefix for the 32-bit Windows API. This works reasonably well for a small number of packages, but the likelihood of name collisions increases as the number of packages grows.

Many programming languages, for example, Java and Ada as well as Modelica, provide a safer and more systematic way of avoiding name collisions through the concept of *package*. A package is simply a container or name space for names of classes, functions, constants, and other allowed definitions. The package name is prefixed to all definitions in the package using standard dot notation. Definitions can be *imported* into the name space of a package.

Modelica has defined the package concept as a restriction and enhancement of the class concept. Thus, inheritance could be used for importing definitions into the name space of another package. However, this gives conceptual modeling problems since inheritance for import is not really a package specialization. Instead, an `import` language construct is provided for Modelica packages. The type name `Voltage` together with all other definitions in `Modelica.SIunits` is imported in the example below, which makes it possible to use it without prefix for declaration of the variable v. By contrast, the declaration of the variable i uses the fully qualified name `Modelica.SIunits.Ampere` of the type `Ampere`, even though the short version also would have been possible. The fully qualified long name for `Ampere` can be used since it is found using the standard nested lookup of the `Modelica` standard library placed in a conceptual top-level package.

```
package MyPack
  import Modelica.SIunits.*;

  class Foo;
    Voltage   v;
    Modelica.SIunits.Ampere   i;
  end Foo;

end MyPack;
```

Importing definitions from one package into another package as in the above example has the drawback that the introduction of new definitions into a package may cause name clashes with definitions in packages using that package. For example, if a definition named v is introduced into the package `Modelica.SIunits`, a compilation error would arise in the package `MyPack`.

An alternative solution to the short-name problem that does not have the drawback of possible compilation errors when new definitions are added to libraries is introducing short convenient name aliases for prefixes instead of long package prefixes. This is possible using the renaming form of `import` statement as in the package `MyPack` below, where the package name SI is introduced instead of the much longer Modelica.SIunits.

Another disadvantage with the above package is that the Ampere type is referred to using standard nested lookup and not via an explicit import statement. Thus, in the worst case we may have to do the following in order to find all such dependencies and the declarations they refer to:

- Visually scan the whole source code of the current package, which might be large.
- Search through all packages containing the current package, that is, higher up in the package hierarchy, since standard nested lookup allows used types and other definitions to be declared anywhere above the current position in the hierarchy.

Instead, a *well-designed package* should state all its dependencies *explicitly* through import statements, which are easy to find. We can create such a package, for example, the package MyPack below, by adding the prefix encapsulated in front of the package keyword. This prevents nested lookup outside the package boundary, ensuring that all dependencies on other packages outside the current package have to be explicitly stated as import statements. This kind of encapsulated package represents an independent unit of code and corresponds more closely to the package concept found in many other programming languages, for example, Java or Ada.

```
encapsulated package MyPack
  import SI = Modelica.SIunits;
  import Modelica;

  class Foo;
    SI.Voltage v;
    Modelica.SIunits.Ampere i;
  end Foo;
  ...

end MyPack;
```

2.17 ANNOTATIONS

A Modelica annotation is extra information associated with a Modelica model. This additional information is used by Modelica environments, for example, for supporting documentation or graphical model editing. Most annotations do not influence the execution of a simulation, that is, the same results should be obtained even if the annotations are removed—but there are exceptions to this rule. The syntax of an annotation is as follows:

annotation(*annotation_elements*)

where *annotation_elements* is a comma-separated list of annotation elements that can be any kind of expression compatible with the Modelica syntax. The following is a resistor class with its associated annotation for the icon representation of the resistor used in the graphical model editor:

```
model Resistor
  ...
  annotation(Icon(coordinateSystem(
   preserveAspectRatio=true,
   extent={{-100,-100},{100,100}}, grid={2,2}),
   graphics={Rectangle(
   extent={{-70,30},{70,-30}},
   lineColor={0,0,255},fillColor={255,255,255},
   fillPattern=FillPattern.Solid),
   Line(points={{-90,0},{-70,0}},
    color={0,0,255}),
  ...
 );
end Resistor;
```

Another example is the predefined annotation choices used to generate menus for the graphical user interface:

```
annotation(choices(choice=1 "P",  choice=2 "PI",
  choice=3 "PID"));
```

The external function annotation arrayLayout can be used to explicitly give the layout of arrays, for example, if it deviates from the defaults rowMajor and columnMajor order for the external languages C and Fortran 77, respectively.

This is one of the rare cases of an annotation influencing the simulation results, since the wrong array layout annotation obviously will have consequences for matrix computations. An example:

```
annotation(arrayLayout = "columnMajor");
```

2.18 NAMING CONVENTIONS

You may have noticed a certain style of naming classes and variables in the examples in this chapter. In fact, certain naming conventions, described below, are being adhered to. These naming conventions have been adopted in the Modelica standard library, making the code more readable and somewhat reducing the risk for name conflicts. The naming conventions are largely followed in the examples in this book and are recommended for Modelica code in general:

- Type and class names (but usually not functions) always start with an uppercase letter, for example, `Voltage`.
- Variable names start with a lowercase letter, for example, `body`, with the exception of some one letter names such as `T` for temperature.
- Names consisting of several words have each word capitalized, with the initial word subject to the above rules, for example, `ElectricCurrent` and `bodyPart`.
- The underscore character is only used at the end of a name, or at the end of a word within a name, to characterize lower or upper indices, for example, `body_low_up`.
- Preferred names for connector instances in (partial) models are `p` and `n` for positive and negative connectors in electrical components, and name variants containing `a` and `b`, for example, `flange_a` and `flange_b`, for other kinds of otherwise identical connectors often occurring in two-sided components.

2.19 MODELICA STANDARD LIBRARIES

Much of the power of modeling with Modelica comes from the ease of reusing model classes. Related classes in particular areas are grouped into packages to make them easier to find.

A special package, called `Modelica`, is a standardized predefined package that together with the Modelica Language is developed and maintained by the Modelica Association. This package is also known as the *Modelica Standard Library*. It provides constants, types, connector classes, partial models, and model classes of components from various application areas, which are grouped into subpackages of the Modelica package, known as the Modelica standard libraries.

The following is a subset of the growing set of Modelica standard libraries currently available:

Modelica.Constants	Common constants from mathematics, physics, and so on
Modelica.Icons	Graphical layout of icon definitions used in several packages
Modelica.Math	Definitions of common mathematical functions
Modelica.SIUnits	Type definitions with SI (international system of units) standard names and units
Modelica.Electrical	Common electrical component models
Modelica.Blocks	Input–output blocks for use in block diagrams
Modelica.Mechanics. Translational	One-dimensional (1D) mechanical translational components
Modelica.Mechanics. Rotational	1D mechanical rotational components
Modelica.Mechanics. MultiBody	MBS library—3D mechanical rigid body multibody models
Modelica.Thermal	Thermal phenomena, heat flow, and other like components
...	...

Additional libraries are available in application areas such as thermodynamics, hydraulics, power systems, data communication, and so forth.

The Modelica Standard Library can be used freely for both noncommercial and commercial purposes under the conditions of *The Modelica License* as stated in the front pages of this book. The full documentation as well as the source code of these libraries appear at the Modelica website.

So far the models presented have been constructed of components from single-application domains. However, one of the main advantages with Modelica is the ease of constructing multidomain models simply by connecting components from different application domain libraries. The DC (direct current) motor depicted in Figure 2.26 is one of the simplest examples illustrating this capability. This

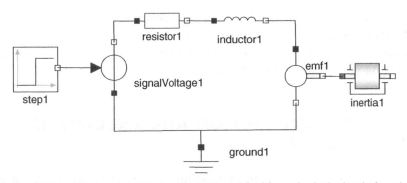

Figure 2.26 Multidomain `DCMotorCircuit` model with mechanical, electrical, and signal block components.

particular model contains components from the three domains, mechanical, electrical, and signal blocks, corresponding to the libraries `Modelica.Mechanics`, `Modelica.Electrical`, and `Modelica.Blocks`.

Model classes from libraries are particularly easy to use and combine when using a graphical model editor, as depicted in Figure 2.27,

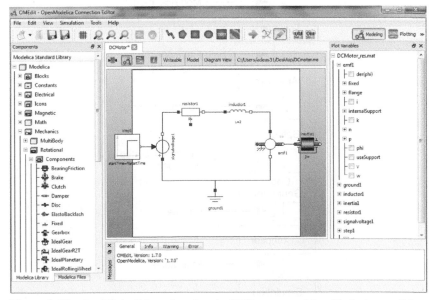

Figure 2.27 Graphical editing of an electrical DC motor model, with the icons of the `Modelica.Mechanics.Rotational` library in the left window.

where the DC motor model is being constructed. The left window shows the `Modelica.Mechanics.Rotational` library, from which icons can be dragged and dropped into the central window when performing graphic design of the model.

2.20 IMPLEMENTATION AND EXECUTION OF MODELICA

In order to gain a better understanding of how Modelica works, it is useful to take a look at the process of translation and execution of a Modelica model, which is sketched in Figure 2.28. First, the Modelica source code is parsed and converted into an internal representation, usually an abstract syntax tree. This representation is analyzed, type checking is done, classes are inherited and expanded, modifications and instantiations are performed, connect equations are converted to

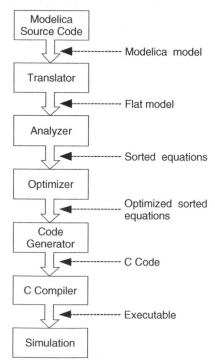

Figure 2.28   Stages of translating and executing a Modelica model.

standard equations, and so forth. The result of this analysis and translation process is a flat set of equations, constants, variables, and function definitions. No trace of the object-oriented structure remains apart from the dot notation within names.

After flattening, all of the equations are topologically sorted according to the data flow dependencies between the equations. In the case of general differential algebraic equations (DAEs), this is not just sorting but also manipulation of the equations to convert the coefficient matrix into block lower triangular form, a so-called BLT transformation. Then an optimizer module containing algebraic simplification algorithms, symbolic index reduction methods, and the like eliminates most equations, keeping only a minimal set that eventually will be solved numerically. As a trivial example, if two syntactically equivalent equations appear, only one copy of the equations is kept. Then independent equations in explicit form are converted to assignment statements. This is possible since the equations have been sorted and an execution order has been established for evaluation of the equations in conjunction with the iteration steps of the numeric solver. If a strongly connected set of equations appears, this set is transformed by a symbolic solver, which performs a number of algebraic transformations to simplify the dependencies between the variables. It can sometimes solve a system of differential equations if it has a symbolic solution. Finally, C code is generated, and linked with a numeric equation solver that solves the remaining, drastically reduced, equation system.

The approximations to initial values are taken from the model definition or are interactively specified by the user. If necessary, the user also specifies the parameter values. A numeric solver for differential algebraic equations (or in simple cases for ordinary differential equations) computes the values of the variables during the specified simulation interval $[t_0, t_1]$. The result of the dynamic system simulation is a set of functions of time, such as R2.v(t) in the simple circuit model. Those functions can be displayed as graphs and/or saved in a file.

In most cases (but not always) the performance of generated simulation code (including the solver) is similar to handwritten C code. Often Modelica is more efficient than straightforwardly written C code because additional opportunities for symbolic optimization are used by the system, compared to what a human programmer can manually handle.

2.20.1 Hand Translation of the Simple Circuit Model

Let us return once more to the simple circuit model, previously depicted in Figure 2.7 but for the reader's convenience also shown below in Figure 2.29. It is instructive to translate this model by hand in order to understand the process.

Classes, instances, and equations are translated into a flat set of equations, constants, and variables (see the equations in Table 2.1), according to the following rules:

1. For each class instance, add one copy of all equations of this instance to the total differential algebraic equation (DAE) system or ordinary differential equation system (ODE)—both alternatives can be possible since a DAE in a number of cases can be transformed into an ODE.

2. For each connection between instances within the model, add connection equations to the DAE system so that nonflow variables are set equal and flow variables are summed to zero.

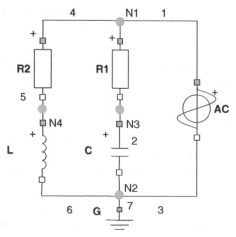

Figure 2.29 SimpleCircuit model once more, with explicitly labeled connection nodes N1, N2, N3, N4, and wires 1 to 7.

Table 2.1

Equations Extracted from Simple Circuit Model — an Implicit DAE System

```
AC  0    = AC.p.i+AC.n.i     L   0   = L.p.i+L.n.i
    AC.v = Ac.p.v-AC.n.v          L.v = L.p.v-L.n.v
    AC.i = AC.p.i                 L.i = L.p.i
    AC.v = AC.VA*                 L.v - L.L*der(L.i)
           sin(2*AC.PI*
             AC.f*time);

R1  0    = R1.p.i+R1.n.i     G   G.p.v = 0
    R1.v = R1.p.v-R1.n.v
    R1.i = R1.p.i
    R1.v = R1.R*R1.i

R2  0    = R2.p.i+R2.n.i   wires R1.p.v = AC.p.v   // wire 1
    R2.v - R2.p.v-R2.n.v         C.p.v  = R1.n.v   // wire 2
    R2.i - R2.p.i               AC.n.v  = C.n.v    // wire 3
    R2.v = R2.R*R2.i            R2.p.v  = R1.p.v   // wire 4
                                L.p.v   = R2.n.v   // wire 5
                                L.n.v   = C.n.v    // wire 6
                                G.p.v   = AC.n.v   // wire 7

C   0    = C.p.i+C.n.i     flow  0 = AC.p.i+R1.p.i+R2.p.i       // N1
    C.v  = C.p.v-C.n.v      at   0 = C.n.i+G.p.i+AC.n.i+L.n.i // N2
    C.i  = C.p.i           node  0 = R1.n.i + C.p.i            // N3
    C.i  = C.C*der(C.v)          0 = R2.n.i + L.p.i            // N4
```

The equation v=p.v-n.v is defined by the class TwoPin. The Resistor class inherits the TwoPin class, including this equation. The SimpleCircuit class contains a variable R1 of type Resistor. Therefore, we include this equation instantiated for R1 as R1.v=R1.p.v-R1.n.v into the system of equations.

The wire labeled 1 is represented in the model as connect (AC.p, R1.p). The variables AC.p and R1.p have type Pin. The variable v is a *nonflow* variable representing voltage potential. Therefore, the equality equation R1.p.v=AC.p.v is generated. Equality equations are always generated when nonflow variables are connected.

Notice that another wire (labeled 4) is attached to the same pin, R1.p. This is represented by an additional connect equation: connect(R1.p.R2.p). The variable i is declared as a flow variable. Thus, the equation AC.p.i+R1.p.i+R2.p.i=0 is

Table 2.2

Variables Extracted from Simple Circuit Model

R1.p.i	R1.n.i	R1.p.v	R1.n.v	R1.v
R1.i	R2.p.i	R2.n.i	R2.p.v	R2.n.v
R2.v	R2.i	C.p.i	C.n.i	C.p.v
C.n.v	C.v	C.i	L.p.i	L.n.i
L.p.v	L.n.v	L.v	L.i	AC.p.i
AC.n.i	AC.p.v	AC.n.v	AC.v	AC.i
G.p.i	G.p.v			

generated. Zero-sum equations are always generated when connecting flow variables, corresponding to Kirchhoff's second law.

The complete set of equations (see Table 2.1) generated from the SimpleCircuit class consists of 32 differential algebraic equations. These include 32 variables, as well as time and several parameters and constants.

Table 2.2 gives the 32 variables in the system of equations, of which 30 are algebraic variables since their derivatives do not appear. Two variables, C.v and L.i, are dynamic variables since their derivatives occur in the equations. In this simple example the dynamic variables are state variables since the DAE reduces to an ODE.

2.20.2 Transformation to State Space Form

The implicit differential algebraic system of equations (DAE system) in Table 2.1 should be further transformed and simplified before applying a numerical solver. The next step is to identify the kind of variables in the DAE system. We have the following four groups:

1. All constant variables, which are model parameters, thus easily modified between simulation runs and declared with the prefixed keyword parameter, are collected into a parameter vector p. All other constants can be replaced by their values, thus disappearing as named constants.

2. Variables declared with the input attribute, that is, prefixed by the input keyword, that appears in instances at the highest hierarchical level, are collected into an input vector u.

3. Variables whose derivatives appear in the model (dynamic variables), that is, the `der()` operator is applied to those variables, are collected into a state vector x.

4. All other variables are collected into a vector y of algebraic variables, that is, their derivatives do not appear in the model.

For our simple circuit model these four groups of variables are the following:

```
p = {R1.R, R2.R, C.C, L.L, AC.VA, AC.f}
u = {AC.v}
x = {C.v,L.i}
y = {R1.p.i, R1.n.i, R1.p.v, R1.n.v, R1.v, R1.i, R2.p.i, R2.n.i,
R2.p.v, R2.n.v, R2.v, R2.i, C.p.i, C.n.i, C.p.v, C.n.v,
  C.i, L.n.i,
L.p.v, L.n.v, L.v, AC.p.i, AC.n.i, AC.p.v, AC.n.v, AC.i, AC.v,
G.p.i, G.p.v}
```

We would like to express the problem as the smallest possible ODE system (in the general case a DAE system) and compute the values of all other variables from the solution of this minimal problem. The system of equations should preferably be in explicit state space form as below.

$$\dot{x} = f(x,t) \tag{2.3}$$

That is, the derivative $\dot{x}$ with respect to time of the state vector x is equal to a function of the state vector x and time. Using an iterative numerical solution method for this ordinary differential equation system, at each iteration step, the derivative of the state vector is computed from the state vector at the current point in time.

For the simple circuit model we have the following:

$$x = \{\texttt{C.v,L.i}\}, \quad u = \{\texttt{AC.v}\}$$
$$\text{(with constants: } \texttt{R1.R,R2.R,C.C,L.L,}$$
$$\dot{x} = \{\textbf{der}(\texttt{C.v}), \textbf{der}(\texttt{L.i})\} \quad \texttt{AC.VA,AC.f,AC.PI)} \tag{2.4}$$

2.20.3 Solution Method

We will use an iterative numerical solution method. First, assume that an estimated value of the state vector x ={C.v,L.i} is available at t=0 when the simulation starts. Use a numerical approximation for the derivative $\dot{x}$ [i.e., der(x)] at time t, for example:

$$\mathbf{der}(x) = (x(t+h) - x(t))/h \qquad (2.5)$$

giving an approximation of x at time t+h:

$$x(t+ h) = x(t) + \mathbf{der}(x)*h \qquad (2.6)$$

In this way the value of the state vector x is computed one step ahead in time for each iteration, provided der(x) can be computed at the current point in simulated time. However, the derivative der(x) of the state vector can be computed from $\dot{x} = f(x,t)$, that is, by selecting the equations involving der(x), and algebraically extracting the variables in the vector x in terms of other variables, as below:

$$\mathbf{der}(C.v) = C.i/C.C$$
$$\mathbf{der}(L.i) = L.v/L.L \qquad (2.7)$$

Other equations in the DAE system are needed to calculate the unknowns C.i and L.v in the above equations. Starting with C.i, using a number of different equations together with simple substitutions and algebraic manipulations, we derive equations (2.8) through (2.10) below.

$$C.i = R1.v/R1.R \qquad (2.8)$$
$$\text{using: } C.i = C.p.i = -R1.n.i = R1.p.i = R1.i$$
$$= R1.v/R1.R$$

$$R1.v = R1.p.v - R1.n.v = R1.p.v - C.v$$
$$\text{using: } R1.n.v = C.p.v = C.v+C.n.v$$
$$= C.v + AC.n.v \qquad (2.9)$$
$$= C.v + G.p.v = C.v + 0 = C.v$$

$$R1.p.v = AC.p.v = AC.VA*sin(2*AC.f*AC.PI*t)$$
$$\text{using: } AC.p.v = AC.v+AC.n.v = AC.v+G.p.v = \qquad (2.10)$$
$$= AC.VA*sin(2*AC.f*AC.PI*t)+0$$

In a similar fashion we derive equations (2.11) and (2.12) below:

$$L.v = L.p.v - L.n.v = R1.p.v - R2.v$$

using: $L.p.v = R2.n.v = R1.p.v- R2.v$ (2.11)

and: $L.n.v = C.n.v = AC.n.v = G.p.v = 0$

$$R2.v = R2.R*L.p.i$$

using: $R2.v = R2.R*R2.i = R2.R*R2.p.i$ (2.12)

$$= R2.R*(-R2.n.i) = R2.R*L.p.i$$

$$= R2.R*L.i$$

Collecting the five equations together:

$$C.i = R1.v/ R1.R$$

$$R1.v = R1.p.v - C.v$$

$$R1.p.v = AC.VA*sin(2*AC.f*AC.PI*time)$$

$$L.v = R1.p.v - R2.v$$

$$R2.v = R2.R*L.i$$ (2.13)

By sorting the equations in data-dependency order, and converting the equations to assignment statements—this is possible since all variable values can now be computed in order—we arrive at the following set of assignment statements to be computed at each iteration, given values of C.v, L.i, and t at the same iteration:

```
R2.v        := R2.R*L.i
R1.p.v      := AC.VA*sin(2*AC.f*AC.PI*time)
L.v         := R1.p.v - R2.v
R1.v        := R1.p.v - C.v
C.i         := R1.v/R1.R
der(L.i)    := L.v/L.L
der(C.v)    := C.i/C.C
```

These assignment statements can be subsequently converted to code in some programming language, for example, C, and executed together with an appropriate ODE solver, usually using better approximations to derivatives and more sophisticated forward-stepping schemes than the simple method described above, which, by the way, is called the *Euler integration* method. The algebraic transformations and sorting procedure that we somewhat painfully performed by hand on the simple circuit example can be performed completely automatically and

	R2.v	R1.p.v	L.v	R1.v	C.i	L.i	C.v
R2.v = R2.R*L.i	1	0	0	0	0	0	0
R1.p.v =	0	1	0	0	0	0	0
AC.VA*sin(2*AC.f*AC.PI*time)							
L.v = R1.p.v - R2.v	1	1	1	0	0	0	0
R1.v = R1.p.v - C.v	0	1	0	1	0	0	0
C.i = R1.v/R1.R	0	0	0	0	1	0	0
der(L.i) = L.v/L.L	0	0	1	0	0	1	0
der(C.v) = C.i/C.C	0	0	0	0	1	0	1

Figure 2.30 Block lower triangular form of the `SimpleCircuit` example.

is known as *BLT transformation*, that is, conversion of the equation system coefficient matrix into block lower triangular form (Fig. 2.30). The remaining 26 algebraic variables in the equation system of the simple circuit model that are not part of the minimal 7-variable kernel ODE system solved above can be computed at leisure for those iterations where their values are desired—this is not necessary for solving the kernel ODE system.

It should be emphasized that the simple circuit example simulated in Fig. 2.31 is trivial. Realistic simulation models often contain tens of thousands of equations, nonlinear equations, hybrid models, and the like. The symbolic transformations and reductions of equation systems performed by a real Modelica compiler are much more complicated than what has been shown in this example, for example, including index reduction of equations and tearing of subsystems of equations, see Fritzson (2004). Index reduction performs symbolic

```
simulate(SimpleCircuit,stopTime=5)]
plot(C.v, xrange={0,5})
```

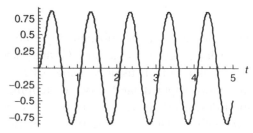

Figure 2.31 Simulation of the `SimpleCircuit` model with a plot of the voltage `C.v` over the capacitor.

2.21 HISTORY

In September 1996, a group of tool designers, application experts, and computer scientists joined forces to work together in the area of object-oriented modeling technology and applications. The group included specialists behind the object-oriented modeling languages Dymola, Omola, ObjectMath, NMF (Neutral Modeling Format), Allan-U.M.L, SIDOPS+, and Smile, even though not all were able to attend the first meeting. Initially, the goal was to write a white paper on object-oriented modeling language technology, including possible unification of existing modeling languages, as part of an action in the ESPRIT project Simulation in Europe Basic Research Working Group (SiE-WG).

However, the work quickly focused on the more ambitious goal of creating a new, unified object-oriented modeling language based on the experience from the previous designs. The designers made an effort to unify the concepts to create a common language, starting the design from scratch. This new language is called *Modelica*.

The group soon established itself as Technical Committee 1 within EuroSim, and as the Technical Chapter on Modelica within the Society for Computer Simulation International (SCS). In February 2000, the Modelica Association was formed as an independent nonprofit international organization for supporting and promoting the development and dissemination of the Modelica Language and the Modelica Standard Libraries.

The first Modelica language description, version 1.0, was put on the Web in September 1997 by a group of proud designers, after a number of intense design meetings. At the date of this writing, the Modelica 3.2 specification is the most recent released–the result of a large amount of work including 34 three-day design meetings. Seven rather complete commercial tools supporting textual and graphical model design and simulation with Modelica are currently available, as well as an almost complete open-source implementation, and several partial university prototypes. A large and growing Modelica Standard Library is also available. The language is quickly spreading both in industry and in academia.

If we trace back a few steps and think about the Modelica technology, two important points become apparent:

- Modelica includes *equations*, which is unusual in most programming languages.
- The Modelica technology includes *graphical* editing for application model design based on predefined components.

In fact, concerning the first point, equations were used very early in human history—already in the third millennium B.C. At that point the well-known equality sign for equations was not yet invented. That happened much later—the equation sign was introduced by Robert Recorde in 1557, in the form depicted in Figure 2.32.

However, it took a while for this invention to spread in Europe. Even a hundred years later, Newton (in his *Principia*, Vol. 1, 1686) still wrote his famous law of motion as text in Latin, as shown in Figure 2.33. Translated to English this can be expressed as: "The change of motion is proportional to the motive force impressed."

In modern mathematical syntax, Newton's law of motion appears as follows:

$$\frac{d}{dt}(m \cdot v) = \sum_i F_i \tag{2.14}$$

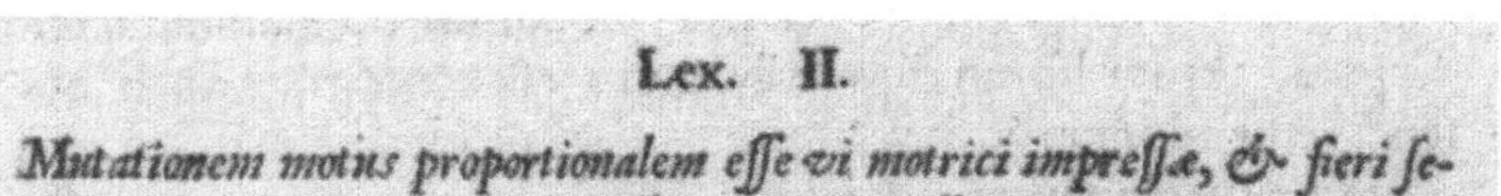

Figure 2.32   Equation sign invented by Robert Recorde 1557. [Reproduced from Figure 4.1-1 on page 81 in (Gottwald et al. (1989); courtesy Thompson Inc.]

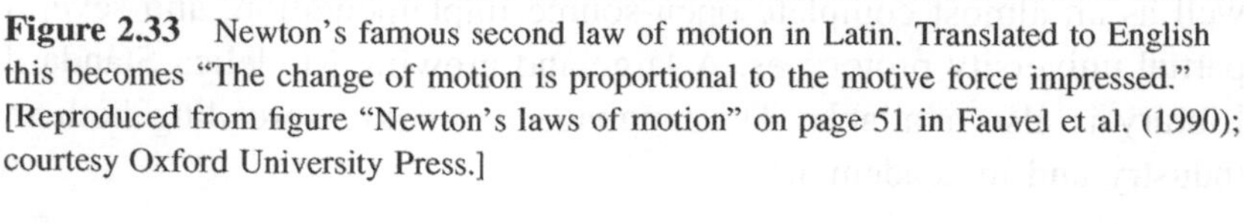

Lex. II.

Mutationem motus proportionalem esse vi motrici impressæ, & fieri secundum lineam rectam qua vis illa imprimitur.

Figure 2.33   Newton's famous second law of motion in Latin. Translated to English this becomes "The change of motion is proportional to the motive force impressed." [Reproduced from figure "Newton's laws of motion" on page 51 in Fauvel et al. (1990); courtesy Oxford University Press.]

This is an example of a differential equation. The first simulators to solve such equations were analog. The idea is to model the system in terms of ordinary differential equations and then make a physical device that obeys the equations. The first analog simulators were mechanical devices, but from the 1950s on electronic analog simulators became predominant. A number of electronic building blocks such as adders, multipliers, integrators, and the like could be interconnected by cables as depicted in Figure 2.34.

Concerning the further development, for a long time equations were quite rare in computer languages. Early versions of Lisp systems and computer algebra systems were being developed, but mostly for formula manipulation rather than for direct simulation.

However, a number of simulation languages for digital computers soon started to appear. The first equation-based modeling tool was Speed-Up, a package for chemical engineering and design introduced in 1964. Somewhat later, in 1967, Simula 67 appeared—the first object-oriented programming language, with profound influence on programming languages and somewhat later on modeling languages. The same year the CSSL (Continuous System Simulation Language)

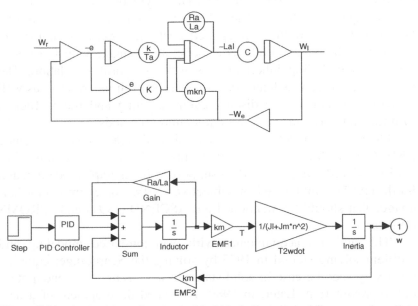

Figure 2.34 Analog computing vs. graphical block diagram modeling on modern digital computers. (Courtesy Karl-Johan ÅAström and Hilding Elmqvist.)

report unified existing notations for expressing continuous system simulation models and introduced a common form of causal "equation," for example:

$$\text{variable} = \text{expression}$$

$$v = \text{INTEG}(F)/m \tag{2.15}$$

The second equation is a variant of the equation of motion: The velocity is the integral of the force divided by the mass. These are not general equations in the mathematical sense since the causality is from right to left, that is, an equation of the form expression = variable is not allowed. However, it is still a definite step forward toward a more general executable computer representation of equation-based mathematical models. ACSL, first introduced in 1976, was a rather common simulation system initially based on the CSSL standard.

An important pioneering predecessor to Modelica is Dymola (*Dynamic modeling language*—not today's Dymola tool, meaning Dynamic modeling laboratory) described in Hilding Elmqvist's Ph.D. Thesis 1978. This was the first work to recognize the importance of modeling with acausal equations together with hierarchical submodels and methods for automatic symbolic manipulation to support equation solution. The GASP-IV system in 1974 followed by GASP-V 1979 introduced integrated continuous-discrete simulation. The Omola language (1989) is a modeling language with full object orientation including inheritance as well as hybrid simulation. The Dymola language was later (1993) enhanced with inheritance, as well as with mechanisms for discrete-event handling and more efficient symbolic-numeric equation system solution methods.

Other early object-oriented acausal modeling languages include NMF (Natural Model Format, 1989), primarily used for building simulation, Allan-U.M.L, SIDOPS+ supporting bond graph modeling, and Smile (1995)—influenced by Objective C. Two other important languages that should be mentioned are ASCEND (1991) and gPROMS (1994).

This author's acquaintance with equation-based modeling and problem solving started in 1975 by solving the Schrödinger equation for a very specific case in solid-state physics, using the pseudopotential approximation. Later, in 1989, I initiated development of a new object-oriented modeling language called ObjectMath together with my brother Dag Fritzson. This was one of the earlier object-oriented computer algebra and simulation systems, integrated with Mathematica and with a general parameterized generic class concept, as well as code generation to C++ for efficient simulation of industrial applications. The fourth version of ObjectMath was completed in the fall of 1996 when I decided to join the Modelica effort instead of continuing with a fifth version of ObjectMath. Later, 1998, we did the first formal executable specification of part of the Modelica language which eventually developed into the OpenModelica open source effort. In December 2007 I initiated the creation of the Open Source Modelica Consortium with initially 7 members, which has expanded to more than 35 members by June 2011.

Concerning the second aspect mentioned earlier—graphical specification of simulation models—Figure 2.34 tells an interesting story. The upper part of Figure 2.34 shows the circuitry of an analog computer with its building blocks connected by cables. The lower part of the figure is a block diagram of very similar structure, directly inspired by the analog computing paradigm but executed on digital computers. Such block diagrams are typically constructed by common tools available today such as Simulink or SystemBuild. Block diagrams represent causal equations since there is a specified data flow direction.

The connection diagrams used in Modelica graphical modeling include connections between instances of classes containing acausal equations, as first explored in the Hibliz system. This is a generalization inspired by the causal analog computing circuit diagrams and block diagrams. The Modelica connection diagrams have the advantage of supporting natural physical modeling since the topology of a connection diagram directly corresponds to the structure and decomposition of the modeled physical system.

2.22 SUMMARY

This chapter has given a quick overview of the most important concepts and language constructs in Modelica. We have also defined important concepts such as object-oriented mathematical modeling and acausal physical modeling and briefly presented the concepts and Modelica language constructs for defining components, connections, and connectors. The chapter concludes with an in-depth example of the translation and execution of a simple model and a short history of equations and mathematical modeling languages up to and including Modelica from ancient times until today.

2.23 LITERATURE

Many programming language books are organized according to a fairly well-established pattern of first presenting a quick overview of the language, followed by a more detailed presentation according to the most important language concepts and syntactic structures.This book is no exception to that rule, where this chapter constitutes the quick overview. As in many other texts we start with a HelloWorld example, for example, as in the Java programming language book (Arnold and Gosling 1999), but with a different contents since printing an "Hello World" message is not very relevant for an equation-based language.

The most important reference document for this chapter is the Modelica tutorial (Modelica Association 2000), of which the first version including a design rationale (Modelica Association 1997) was edited primarily by Hilding Elmqvist. Several examples, code fragments, and text fragments in this chapter are based on similar ones in the tutorial, for example, the SimpleCircuit model with the simple electrical components, the polynomialEvaluator, the low-pass filter, the ideal Diode, and the BouncingBall model. Figure 2.8 on block-oriented modeling is also from the tutorial. Another important reference document for this chapter is the Modelica language specification (Modelica Association 2010) Some formulations from the Modelica specification regarding operator overloading and stream connectors are reused in this chapter in order to state the same semantics.

The hand translation of the simple circuit model is inspired by a similar but less elaborated example in a series of articles by Martin Otter et al. (1999). The recent history of mathematical modeling languages is described in some detail in Åström et al. (1998), whereas bits and pieces of the ancient human history of the invention and use of equations can be found in Gottwald et al. (1989), and the picture on Newton's second law in Latin in Fauvel et al. (1990). Early work (GASP-IV) on combined continuous/discrete simulation is described Pritsker (1974) followed by Cellier (1979) in the GASP-V system. This author's first simulation work involving solution of the Schrödinger equation for a particular case is described in Fritzson and Berggren (1976).

The predecessors of the Modelica language are briefly described in Appendix F, including Dymola meaning the Dynamic Modeling Language: (Elmqvist 1978; Elmqvist et al. 1996), Omola: (Mattsson et al. 1993; Andersson 1994), ObjectMath: (Fritzson et al. 1992, 1995; Viklund and Fritzson 1995), NMF: (Sahlin et al. 1996), and Smile: (Ernst et al. 1997).

Speed-Up, the earliest equation-based simulation tool, is presented in Sargent and Westerberg (1964), whereas Simula-67—the first object-oriented programming language—is described in Birtwistle et al. (1974). The early CSSL language specification is described in Augustin et al. (1967), whereas the ACSL system is described in Mitchell and Gauthier (1986). The Hibliz system for a hierarchical graphical approach to modeling is presented in Elmqvist and Mattsson (1982, and 1989).

Software component systems are presented in Assmann (2002) and Szyperski (1997).

The Simulink system for block-oriented modeling is described in MathWorks (2001), whereas the MATLAB language and tool are described in MathWorks (2002).

The DrModelica electronic notebook with the examples and exercises of this book has been inspired by DrScheme (Felleisen et al. 1998) and DrJava (Allen et al. 2002), as well as by Mathematica (Wolfram 1997), a related electronic book for teaching mathematics (Davis et al. 1994), and the MathModelica environment (Fritzson, Engelson and Gunnarsson 1998; Fritzson, Gunnarsson and Jirstrand 2002). The first version of DrModelica is described in Lengquist-Sandelin and Monemar (2003a, 2003b).

There are general Modelica articles and books (Elmqvist and Mattsson 1997; Fritzson and Engelson 1998; Elmqvist et al. 1999), a series of 17 articles (in German) of which Otter (1999) is the first, (Tiller 2001; Fritzson and Bunus 2002; Elmqvist et al. 2002; Fritzson 2004).

The proceedings from the following conferences, as well as some not listed here, contain a number of Modelica-related papers: the Scandinavian Simulation Conference (Fritzson 1999), and especially the International Modelica Conferences (Fritzson 2000; Otter 2002; Fritzson 2003, Schmitz 2005, Kral and Haumer 2006; Bachmann 2008; Casella 2009; Clauß 2011).

2.24 EXERCISES

2.1. What is a class?

Creating a Class: Create a class, Add, that calculates the sum of two parameters, which are Integer numbers with given values.

2.2. What is an instance?

Creating Instances:

```
class Dog
  constant Real legs = 4;
  parameter String name = "Dummy";
end dog;
```

- Create an instance of the class Dog.
- Create another instance and give the dog the name "Tim".

2.3. Write a function, average, that returns the average of two Real values. Make a function call to average with the input 4 and 6.

2.4. What do the terms partial, class, and extends stand for?

2.5. *Inheritance:* Consider the Bicycle class below.

```
record Bicycle
  Boolean has_wheels = true;
  Integer nrOfWheels = 2;
end Bicycle;
```

Define a record, ChildrensBike, that inherits from the class Bicycle and is meant for kids. Give the variables values.

2.6. *Declaration Equations and Normal Equations:* Write a class, Birthyear, which calculates the year of birth from this year together with a person's age. Point out the declaration equations and the normal equations.

Modification Equation: Write an instance of the class Birthyear above. The class, let's call it MartinsBirthyear, will calculate Martin's year of birth, call the variable martinsBirthyear, who is 29 years old. Point out the modification equation.

Check your answer, for example, by writing as below.[8]
val(martinsBirthday.birthYear, 0)

---

[8]Using the OpenModelica command line interface or OMNotebook commands, the expression val(martinsBirthday.birthYear,0) means the birthYear value at time=0, at the beginning of the simulation. It is also in many cases possible to interactively enter an expression such as martinsBirthday.birthYear and get back the result without giving the time argument.

2.7. *Classes:*

```
class Ptest
  parameter Real x;
  parameter Real y;
  Real z;
  Real w;
equation
  x + y = z;
end Ptest;
```

What is wrong with this class? Is there something missing?

2.8. Create a record containing several vectors and matrices:

- A vector containing the two `Boolean` values `true` and `false`
- A vector with five `Integer` values of your choice
- A matrix of three rows and four columns containing `String` values of your choice
- A matrix of one row and five columns containing different `Real` values, also those of your choice

2.9. Can you really put an algorithm section inside an equation section?

2.10. *Writing an Algorithm Section:* Create the class, `Average`, which calculates the average between two integers, using an algorithm section. Make an instance of the class and send in some values.

Simulate and then test the result of the instance class by writing `instanceVariable.classVariable`.

2.11. (A harder exercise) Write a class, `AverageExtended`, that calculates the average of four variables (`a`, `b`, `c`, and `d`). Make an instance of the class and send in some values.

Simulate and then test the result of the instance class by writing `instanceVariable.classVariable`.

2.12. *If equation:* Write a class `Lights` that sets the variable switch (integer) to one if the lights are on and zero if the lights are off.

When equation: Write a class `LightSwitch` that is initially switched off and switched on at time 5.

Tip: `sample(start, interval)` returns true and triggers time events at time instants and `rem(x, y)` returns the integer remainder of x/y such that `div(x,y) * y + rem(x, y) = x`.

2.13. What is a package?

Creating a Package: Create a package that contains a division function (that divides two `Real` numbers) and a constant `k = 5`.

Create a class, containing a variable x. The variable gets its value from the division function inside the package, which divides 10 by 5.

CHAPTER 3

Classes and Inheritance

The fundamental unit of modeling in Modelica is the class. Classes provide the structure for objects, also known as instances, and serve as templates for creating objects from class definitions. Classes can contain equations, which provide the basis for the executable code that is used for computation in Modelica. Conventional algorithmic code can also be part of classes. Interaction between objects of well-structured classes in Modelica is usually done through so-called connectors, which can be seen as "access ports" to objects. All data objects in Modelica are instantiated from classes, including the basic data types—Real, Integer, String, Boolean—and enumeration types, which are built-in classes or class schemata.

A class in Modelica is essentially equivalent to a type. Declarations are the syntactic constructs needed to introduce classes and objects.

3.1 CONTRACT BETWEEN CLASS DESIGNER AND USER

Object-oriented modeling languages try to separate the notion of *what* an object is from *how* its behavior is implemented and specified in detail. The "what" of an object in Modelica is usually described through documentation including graphics and icons, together with possible public connectors, variables, and other public elements, and

Introduction to Modeling and Simulation of Technical and Physical Systems with Modelica,
First Edition. By Peter Fritzson

their associated semantics. For example, the what of an object of class `Resistor` is the documentation that it models a "realistic" ideal resistor coupled with the fact that its interaction with the outside world is through two connectors `n, p` of type `Pin`, and their semantics. This combination—documentation, connectors and other public elements, and semantics—is often described as a *contract* between the designer of the class and the modeler who uses it, since the what part of the contract specifies to the modeler what a class represents, whereas the "how" provided by the class designer implements the required properties and behavior.

An incorrect, but common, assumption is that the connectors and other public elements of a class (its "signature") specify its entire contract. This is not correct since the intended semantics of the class is also part of the contract even though it might only be publicly described in the documentation, and internally modeled through equations and algorithms. Two classes, for example, a `Resistor` and a temperature-dependent `Resistor`, may have the same signature in terms of connectors but still not be equivalent since they have different semantics. The contract of a class includes both the signature and the appropriate part of its semantics together.

The how of an object is defined by its class. The implementation of the behavior of the class is defined in terms of equations and possibly algorithmic code. Each object is an instance of a class. Many objects are composite objects that is, consist of instances of other classes.

3.2 A CLASS EXAMPLE

The basic properties of a class are given by:

- Data contained in variables declared in the class
- Behavior specified by equations together with possible algorithms

Here is a simple class called `CelestialBody` that can be used to store data related to celestial bodies such as the Earth, the moon, asteroids, planets, comets, and stars:

```
class CelestialBody
  constant  Real   g = 6.672e-11;
```

```
  parameter Real   radius;
  parameter String name;
  Real             mass;
end CelestialBody;
```

The declaration of a class starts with a keyword such as class or model, followed by the name of the class. A class declaration creates a *type name* in Modelica, which makes it possible to declare variables of that type, also known as objects or instances of that class, simply by prefixing the type name to a variable name:

```
CelestialBody moon;
```

This declaration states that moon is a variable containing an object of type CelestialBody. The declaration actually creates the object, that is, allocates memory for the object. This is in contrast to a language like Java, where an object declaration just creates a reference to an object.

This first version of CelestialBody is not very well designed. This is intentional, since we will demonstrate the value of certain language features for improving the class in this and the following two chapters.

3.3 VARIABLES

The variables belonging to a class are sometimes called record fields or attributes; the CelestialBody variables radius, name, and mass are examples. Every object of type CelestialBody has its own instances of these variables. Since each separate object contains a different instance of the variables, this means that each object has its own unique state. Changing the mass variable in one CelestialBody object does not affect the mass variables in other CelestialBody objects.

Certain programming languages, for example, Java and C++, allow so-called static variables, also called class variables. Such variables are shared by all instances of a class. However, this kind of variable is not available in Modelica.

A declaration of an instance of a class, for example, moon being an instance of CelestialBody, allocates memory for the object and initializes its variables to appropriate values. Three of the variables in the class `CelestialBody` have special status: the gravitational *constant* g is a `constant` that never changes and can be substituted by its value. The *simulation parameters* `radius` and name are examples of a special kind of "constant," denoted by the keyword `parameter` in Modelica. Such simulation parameter constants are assigned their values only at the start of the simulation and keep their values constant during simulation.

In Modelica variables store results of computations performed when solving the equations of a class together with equations from other classes. During solution of time-dependent problems, the variables store results of the solution process at the current time instant.

As the reader may have noted, we use the terms *object* and *instance* interchangeably with the same meaning; we also use the terms *record field, attribute*, and *variable* interchangeably. Sometimes the term *variable* is used interchangeably with *instance* or *object*, since a variable in Modelica always contains an instance of some class.

3.3.1 Duplicate Variable Names

Duplicate variable names are not allowed in class declarations. The name of a declared element, for example, a variable or local class, must be different from the names of all other declared elements in the class. For example, the following class is illegal:

```
class IllegalDuplicate
  Real    duplicate;
  Integer duplicate;
    // Error! Illegal duplicate variable name
end IllegalDuplicate;
```

3.3.2 Identical Variable Names and Type Names

The name of a variable is not allowed to be identical to its type specifier. Consider the following erroneous class:

```
class IllegalTypeAsVariable
  Voltage Voltage;
    // Error! Variable name must be different from type
  Voltage voltage;
    // Ok! Voltage and voltage are different names
end IllegalTypeAsVariable;
```

The first variable declaration is illegal since the variable name is identical to the type specifier of the declaration. The reason this is a problem is that the `Voltage` type lookup from the second declaration would be hidden by a variable with the same name. The second variable declaration is legal since the lowercase variable name `voltage` is different from its uppercase type name `Voltage`.

3.3.3 Initialization of Variables

The default-suggested (the solver may choose otherwise, if not `fixed`) initial variable values are the following, if no explicit start values are specified (see Section 2.3.2):

- The value zero as the default initial value for numeric variables
- The empty string `""` for `String` variables
- The value `false` for `Boolean` variables
- The lowest enumeration value in an enumeration type for enumeration variables

However, *local variables* to functions have *unspecified* initial values if no defaults are explicitly given. Initial values can be explicitly specified by setting the `start` attributes of instance variables equal to some value, or providing initializing assignments when the instances are local variables or formal parameters in functions. For example, explicit start values are specified in the class `Rocket` shown in the next section for the variables `mass`, `altitude`, and `velocity`.

3.4 BEHAVIOR AS EQUATIONS

Equations are the primary means of specifying the behavior of a class in Modelica, even though algorithms and functions also are available. The way in which the equations interact with equations from other classes determines the solution process, that is, program execution, where successive values of variables are computed over time. This is exactly what happens during dynamic system simulation. During solution of time-dependent problems, the variables store results of the solution process at the current time instant.

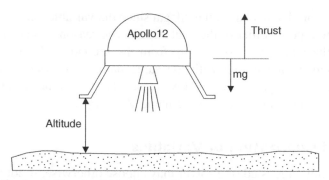

Figure 3.1 Apollo12 rocket for landing on the moon.

The class Rocket embodies the equations of vertical motion for a rocket (e.g., as depicted in Fig. 3.1), which is influenced by an external gravitational force field gravity , and the force thrust from the rocket motor, acting in the opposite direction to the gravitational force, as in the expression for acceleration below:

$$\text{acceleration} = \frac{\text{thrust} - \text{mass} \times \text{gravity}}{\text{mass}}$$

The following three equations are first-order differential equations stating well-known laws of motion between altitude, vertical velocity, and acceleration:

$$\text{mass}' = -\text{massLossRate} \cdot \text{abs(thrust)}$$

$$\text{altitude}' = \text{velocity}$$

$$\text{velocity}' = \text{acceleration}$$

All these equations appear in the class Rocket below, where the mathematical notation (′) for derivative has been replaced by the pseudofunction der() in Modelica. The derivative of the rocket mass is negative since the rocket fuel mass is proportional to the amount of thrust from the rocket motor.

```
class Rocket "rocket class"
  parameter String name;
  Real mass(start=1038.358);
  Real altitude(start= 59404);
  Real velocity(start= -2003);
  Real acceleration;
  Real thrust;    // Thrust force on the rocket
```

```
   Real gravity;    // Gravity force field
   parameter Real massLossRate=0.000277;
 equation
   (thrust-mass*gravity)/mass = acceleration;
   der(mass)  = -massLossRate * abs(thrust);
   der(altitude) = velocity;
   der(velocity) = acceleration;
 end Rocket;
```

The following equation, specifying the strength of the gravitational force field, is placed in the class MoonLanding in the next section since it depends on both the mass of the rocket and the mass of the moon:

$$\text{gravity} = \frac{g_{\text{moon}} \cdot \text{mass}_{\text{moon}}}{(\text{altitude}_{\text{apollo}} + \text{radius}_{\text{moon}})^2}$$

The amount of thrust to be applied by the rocket motor is specific to a particular class of landings and, therefore, also belongs to the class MoonLanding:

```
thrust = if (time < thrustDecreaseTime)then
            force1
        else if (time < thrustEndTime)then
            force2
        else 0
```

3.5 ACCESS CONTROL

Members of a Modelica class can have two levels of visibility: public or protected . The default is public if nothing else is specified, for example, regarding the variables force1 and force2 in the class MoonLanding below. The public declaration of force1, force2, apollo, and moon means that any code with access to a MoonLanding instance can read or update those values.

The other possible level of visibility, specified by the keyword protected—for example, for the variables thrustEndTime and thrustDecreaseTime, means that only code *inside* the class as well as code in classes that inherit this class are allowed access. Code inside a class includes code from local classes. However, only code inside the class is allowed access to the *same* instance of a protected variable—classes that extend the class will naturally access another instance of a protected variable since declarations are "copied" at inheritance. This is different from corresponding access rules for Java.

Note that an occurrence of one of the keywords public or protected means that all member declarations following that keyword assume the corresponding visibility until another occurrence of one of those keywords.

The variables thrust gravity, and altitude belong to the apollo instance of the Rocket class and are therefore prefixed by apollo in references such as apollo.thrust. The gravitational constant g, the mass, and the radius belong to the particular celestial body called moon on which surface the apollo rocket is landing.

```
class MoonLanding
  parameter Real force1 = 36350;
  parameter Real force2 = 1308;
  protected
  parameter Real thrustEndTime = 210;
  parameter Real thrustDecreaseTime = 43.2;
public
  Rocket          apollo(name="apollo12");
  CelestialBody   moon(name="moon",mass=7.382e22,radius=1.738e6);
  equation
  apollo.thrust = if (time<thrustDecreaseTime) then force1
                  else if (time<thrustEndTime) then force2
                  else 0;
  apollo.gravity = moon.g * moon.mass /
    (apollo.altitude + moon.radius)^2;
end MoonLanding;
```

3.6 SIMULATING THE MOON LANDING EXAMPLE

We simulate the MoonLanding model during the time interval {0,230} by the following command, using the OpenModelica simulation environment:

```
simulate(MoonLanding, stopTime=230)
```

Since the solution for the altitude of the Apollo rocket is a function of time, it can be plotted in a diagram (Fig. 3.2). It starts at an altitude of 59,404 m (not shown in the diagram) at time zero, gradually reducing it until touchdown at the lunar surface when the altitude is zero.

```
plot(apollo.altitude, xrange={0,208})
```

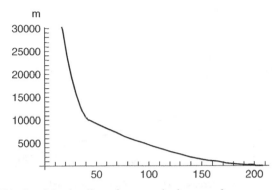

Figure 3.2 Altitude of the Apollo rocket over the lunar surface.

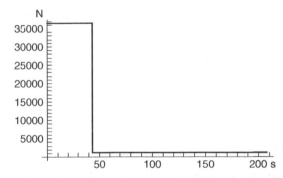

Figure 3.3 Thrust from the rocket motor, with an initial high thrust f1 followed by a lower thrust f2.

The thrust force from the rocket is initially high but is reduced to a low level after 43.2 s, that is, the value of the simulation parameter thrustDecreaseTime, as shown in Figure 3.3.

```
plot(apollo.thrust, xrange={0,208})
```

The mass of the rocket decreases from initially 1038.358 kg to around 540 kg as the fuel is consumed (Fig. 3.4).

```
plot(apollo.mass, xrange={0,208})
```

The gravity field increases when the rocket gets closer to the lunar surface, as depicted in Figure 3.5, where the gravity has increased to 1.63 N/kg after 200 s.

```
plot(apollo.gravity, xrange={0,208})
```

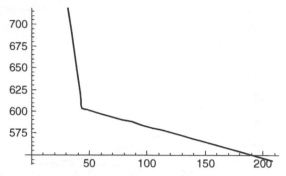

Figure 3.4   Rocket mass decreases when the fuel is consumed.

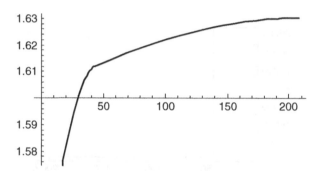

Figure 3.5   Gradually increasing gravity when the rocket approaches the lunar surface.

The rocket initially has a high negative velocity when approaching the lunar surface. This is reduced to zero at touchdown, giving a smooth landing, as shown in Figure 3.6.

```
plot(apollo.velocity, xrange={0,208})
```

When experimenting with the MoonLanding model, the reader might notice that the model is nonphysical, regarding at least one important aspect of the landing. After touchdown when the speed has been reduced to zero, if the simulation is allowed to continue, the speed will increase again and the lander will accelerate toward the center of the moon. This is because we have left out the ground contact force from the lunar surface acting on the lander after it has landed, which will prevent this from happening. It is left as an exercise to introduce such a ground force into the MoonLanding model.

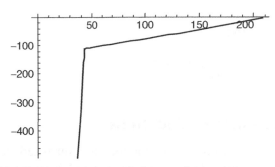

Figure 3.6   Vertical velocity relative to the lunar surface.

3.7 INHERITANCE

Let us regard an example of extending a very simple Modelica class, for example, the class `ColorData` introduced in Section 2.4. We show two classes named `ColorData` and `Color`, where the derived class (subclass) `Color` inherits the variables to represent the color from the *base class* (superclass) `ColorData` and adds an equation as a constraint on the color values.

```
record ColorData
   Real    red;
   Real    blue;
   Real    green;
end ColorData;

class Color
   extends ColorData;
equation
   red + blue + green = 1;
end Color;
```

In the process of inheritance, the data and behavior of the super-class in the form of variable and attribute declarations, equations, and certain other contents are copied into the subclass. However, as we already mentioned, before the copying is done certain type expansion, checking, and modification operations are performed on the inherited definitions. The expanded `Color` class is equivalent to the following class:

```
class ExpandedColor
   Real red;
```

```
    Real blue;
    Real green;
equation
    red + blue + green = 1;
end ExpandedColor;
```

3.7.1 Inheritance of Equations

In the previous section we mentioned that inherited equations are copied from the superclass or base class and inserted into the subclass or derived class. What happens if there already is an *identical equation* locally declared in the derived class? In that case there will be two identical equations, making the system overdetermined and impossible to solve.

```
class Color2
   extends Color;
equation
   red + blue + green = 1;
end Color2;
```

The expanded Color2 class is equivalent to the following class:

```
class ExpandedColor2
   Real red;
   Real blue;
   Real green;
equation
   red + blue + green = 1;
   red + blue + green = 1;
end ExpandedColor2;
```

3.7.2 Multiple Inheritance

Multiple inheritance, that is, several extends statements, is supported in Modelica. This is useful when a class wishes to include several orthogonal kinds of behavior and data, for example, combining geometry and color.

For example, the new class ColoredPoint inherits from multiple classes, that is, uses multiple inheritance, to get the *position* variables from class Point, as well as the color variables together with the equation from class Color.

```
class Point
   Real x;
   Real y, z;
end Point;

class ColoredPoint
   extends Point;
   extends Color;
end ColoredPoint;
```

In many object-oriented programming languages multiple inheritance causes problems when the same definition is inherited twice, but through different intermediate classes. A well-known case is the so-called diamond inheritance (Fig. 3.7):

The class Point contains a coordinate position defined by the variables x and y. The class VerticalLine inherits Point but also adds the variable vlength for the line length. Analogously, the class HorizontalLine inherits the position variables x, y and adds a horizontal length. Finally, the class Rectangle is intended to have position variables x, y, a vertical length, and a horizontal length.

```
class Point
   Real x;
   Real y;
end Point;

class VerticalLine
   extends Point;
   Real vlength;
end VerticalLine;

class HorizontalLine
   extends Point;
   Real hlength;
end HorizontalLine;
```

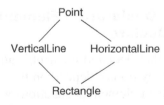

Figure 3.7 Diamond inheritance.

```
class Rectangle
  extends VerticalLine;
  extends HorizontalLine;
end Rectangle;
```

The potential problem is that we have diamond inheritance since the coordinate position defined by the variables x and y is inherited twice: both from VerticalLine and from HorizontalLine. Should the position variables from VerticalLine be used or the ones from HorizontalLine? Is there some way to resolve the problem?

In fact, there is a way. In Modelica diamond inheritance is not a problem since there is a rule stating that if several identical declarations or equations are inherited, only one of them is kept. Thus, there will be only one set of position variables in the class Rectangle, making the total set of variables in the class the following: x, y, vlength, and hlength. The same is true for the classes Rectangle2 and Rectangle3 below.

```
class Rectangle2
  extends Point;
  extends VerticalLine;
  extends HorizontalLine;
end Rectangle;
```

```
class Rectangle3
  Real x, y;
  extends VerticalLine;
  extends HorizontalLine;
end Rectangle;
```

The reader might perhaps think that there could be cases where the result depends on the relative order of the extends clauses. However, this is in fact not possible in Modelica, which we shall see in Section 3.7.4.

3.7.3 Processing Declaration Elements and Use Before Declare

In order to guarantee that declared elements can be used before they are declared and that they do not depend on their order of declaration, the lookup and analysis of element declarations within a class proceeds as follows:

1. The *names* of declared local classes, variables, and other attributes are found. Also, modifiers are merged with the local element declarations, and redeclarations are applied.

2. *Extends clauses* are processed by lookup and expansion of inherited classes. Their contents is expanded and inserted into the current class. The lookup of the inherited classes should be *independent*, that is, the analysis and expansion of one extends clause should not be dependent on another.

3. All element declarations are expanded and type checked.

4. A check is performed that all elements with the same name are identical.

The reason that all the names of local types, variables, and other attributes need to be found first is that a *use* of an element can come before its *declaration*. Therefore, the names of the elements of a class need to be known before further expansion and type analysis is performed.

For example, the classes Voltage and Lpin are used before they are declared within the class C2:

```
class C2
   Voltage v1, v2;
   Lpin    pn;

   class Lpin
     Real p;
   end Lpin;

   class Voltage = Real(unit="kV");
end C2;
```

3.7.4 Declaration Order of extends Clauses

We have already stated in Chapter 2 and in the previous section that in Modelica the *use* of declared items is independent of the order in which they are declared, except for function formal parameters and record fields (variables). Thus, variables and classes can be used before they are declared. This also applies to extends clauses. The order in which extends clauses are stated within a class does not matter with regard to declarations and equations inherited via those extends clauses.

3.7.5 The MoonLanding **Example Using Inheritance**

In the MoonLanding example from Section 3.4 the declaration of certain variables like mass and name were repeated in each of the classes CelestialBody and Rocket. This can be avoided by collecting those variable declarations into a generic body class called Body, and reusing these by inheriting Body into CelestialBody and Rocket. This restructured MoonLanding example appears below. We have replaced the general keyword class by the specialized class keyword model, which has the same semantics as class apart from the fact that it cannot be used in connections. The reason is that it is more common practice to use the keyword model for modeling purposes than to use the keyword class. The first model is the generic Body class designed to be inherited by more specialized kinds of bodies.

```
model Body   "generic body"
   Real    mass;
   String name;
end Body;
```

The CelestialBody class inherits the generic Body class and is in fact a specialized version of Body. Compare with the version of CelestialBody without inheritance previously presented in Section 3.4.

```
model CelestialBody "celestial body"
   extends Body;
   constant  Real g = 6.672e-11;
   parameter Real radius;
end CelestialBody;
```

The Rocket class also inherits the generic Body class and can, of course, be regarded as another specialized version of Body.

```
model Rocket "generic rocket class"
   extends Body;
   parameter Real massLossRate=0.000277;
   Real altitude(start= 59404);
   Real velocity(start= -2003);
   Real acceleration;
   Real thrust;
   Real gravity;
equation
```

```
    thrust - mass * gravity = mass * acceleration;
    der(mass)   = -massLossRate * abs(thrust);
    der(altitude) = velocity;
    der(velocity) = acceleration;
  end Rocket;
```

The MoonLanding class below is identical to the one presented in Section 3.4, apart from the change of keyword from class to model.

```
model MoonLanding
  parameter Real force1 = 36350;
  parameter Real force2 = 1308;
  parameter Real thrustEndTime = 210;
  parameter Real thrustDecreaseTime = 43.2;
  Rocket     apollo(name="apollo12", mass(start=1038.358) );
  CelestialBody moon(mass=7.382e22,radius=1.738e6,name="moon");
equation
  apollo.thrust - if (time<thrustDecreaseTime) then force1
                  else if (time<thrustEndTime) then force2
                  else 0;
  apollo.gravity = moon.g*moon.mass
/(apollo.altitude+moon.radius)^2;
end Landing;
```

We simulate the restructured MoonLanding model during the time interval {0, 230}:

```
simulate(MoonLanding, stopTime=230)
```

The result is identical to the simulation of MoonLanding in Section 3.5, which it should be. See, for example, the altitude of the Apollo rocket as a function of time, plotted in Figure 3.8.

```
plot(apollo.altitude, xrange={0,208})
```

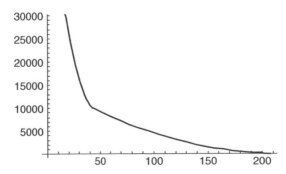

Figure 3.8 Altitude of the Apollo rocket over the lunar surface.

3.8 SUMMARY

This chapter has its focus on the most important structuring concept in Modelica: the class concept. We started by a tutorial introduction of the idea of contract between designer and user together with the basics of Modelica classes as exemplified by the moon landing model.

3.9 LITERATURE

An important reference document for this chapter is the Modelica language specification Modelica Association (2010). Several examples in this chapter are based on similar examples in that document and in Modelica Association (2000). In the beginning of this chapter we mention the idea of a contract between software designers and users. This is part of methods and principles for object-oriented modeling and design of software described in Rumbaugh et al. (1991), Booch (1991, 1994), and Meyer (1997). The moon landing example is based on the same formulation of Newton's equations as in the lunar landing example in Cellier (1991).

CHAPTER 4

System Modeling Methodology

So far in this text we have primarily discussed the principles of object-oriented mathematical modeling, a number of Modelica language constructs to support high-level model representation and a high degree of model reuse, and many model examples that demonstrate the use of these language constructs.

However, we have not yet presented any systematic method on how to create a model of a system to be investigated. Nor have we previously presented the underlying mathematical state space equations in any detail. These are the topics of the current chapter. However, the state space representation presented here covers only the simple case for continuous systems.

4.1 BUILDING SYSTEM MODELS

A basic question is: How do we arrive at reasonable mathematical models of the systems we want to study and possibly simulate? That is, what is an effective modeling process?

The application domain is of primary importance in all kinds of modeling. We talk about *physical modeling* when the systems to be modeled are described by natural laws of physics, chemistry, biology, mechanical and electrical engineering, and the like, and these laws can

Introduction to Modeling and Simulation of Technical and Physical Systems with Modelica,
First Edition. By Peter Fritzson
© 2011 the Institute of Electrical and Electronics Engineers, Inc. Published 2011 by John Wiley & Sons, Inc.

be represented directly in the mathematical model we are constructing. However, it does not really matter whether the application domain is "physical" or not. There are laws governing economic systems, data communication, information processing, and so forth that should be more or less directly expressible in the model we are building when using a high-level modeling approach. All of these laws are given by the universe where we exist (but formulated by human beings) and can be regarded as basic laws of nature.

At the start of a modeling effort we first need to identify which application domains are involved regarding the system we want to model, and for each of these domains find out the relevant governing laws that influence the phenomena we want to study.

In order to be able to handle the complexity of large models and to reduce the effort of model building by reusing model components, it is quite useful to apply hierarchical decomposition and object-oriented component-based techniques such as those advocated in this text. To make this more clear, we will briefly contrast the traditional physical modeling approach to the object-oriented component-based approach.

However, one should be aware that even the traditional approach to physical modeling is "higher level" than certain other approaches such as block-oriented modeling or direct programming in common imperative languages, where the user has to manually convert equations into assignment statements or blocks and manually restructure the code to fit the data and signal flow context for a specific model use.

4.1.1 Deductive Modeling Versus Inductive Modeling

So far we have dealt almost exclusively with the so-called *deductive modeling* approach, also called *physical modeling* approach, where the behavior of a system is *deduced* from the application of natural laws expressed in a model of the system. Such models are created based on an understanding of the "physical" or "artificial" processes underlying the system in question, which is the basis for the term "physical modeling."

However, in many cases, especially for biological and economic systems, accurate knowledge about complex systems and their internal processes may not be available to an extent that would be needed to support physical modeling. In such application areas it is common to use an entirely different approach to modeling. You make observations about the system under study and then try to fit a hypothetical mathematical model to the observed data by adapting the model, typically by finding values of unknown coefficients. This is called the *inductive modeling* approach.

Inductive models are directly based on measured values. This makes such models difficult to validate beyond the observed values. For example, we would like to accommodate mechanisms in a system model is such a way that disasters can be predicted and therefore possibly prevented. However, this might be impossible without first observing a real disaster event (that we would like to avoid by all means) in the system. This is a clear disadvantage of inductive models.

Also notice that just by adding sufficiently many parameters to an inductive model it is possible to fit virtually any model structure to virtually any data. This is one of the most severe disadvantages of inductive models since in this way one can be easily fooled concerning their validity.

In the rest of this book we will mainly deal with the deductive or physical modeling approach, except for a few biological application examples where we present models that are partly inductive and partly have some physical motivation.

4.1.2 Traditional Approach

The traditional methodology for physical modeling can be roughly divided into three phases:

1. Basic structuring in terms of variables
2. Stating equations and functions
3. Converting the model to state space form

The first phase involves identification of which variables are of interest, for example, for the intended use of the model, and the roles of these variables. Which variables should be considered as *inputs*, for example, external signals, as *outputs*, or as internal *state* variables?

Which quantities are especially important for describing what happens in the system? Which are time varying and which are approximately *constant*? Which variables influence other variables?

The second phase involves stating the basic equations and formulas of the model. Here we have to look for the governing laws of the application domains that are relevant for the model, for example, *conservation equations* of quantities of the *same kind* such as power in related to power out, input flow rate related to output flow rates, and conservation of quantities such as energy, mass, charge, information, and the like. Formulate *constitutive equations* that relate quantities of *different kind*, for example, relating voltage and current for a resistor, input and output flows for a tank, input and output packets for a communication link, and so forth. Also formulate relations, for example, involving material properties and other system properties. Consider the level of precision and appropriate approximation trade-offs in the relationships.

The third phase involves converting the model in its current form consisting of a set of variables and a set of equations to a *state space* equation system representation that fits the numeric solver to be used. A set of state variables needs to be chosen, their time derivatives (for dynamic variables) expressed as functions of state variables and input variables (for the case of state space equations in explicit form), and the output variables expressed as functions of state variables and input variables. If a few unnecessary state variables are included, this causes no harm, just some unnecessary computation. This phase is completely eliminated when using Modelica since the conversion to state space form is performed automatically by the Modelica compiler.

4.1.3 Object-Oriented Component-Based Approach

When using the object-oriented component-based approach to modeling we first try to understand the system structure and decomposition in a hierarchical top-down manner. When the system components and interactions between these components have been roughly identified, we can apply the first two traditional modeling phases of identifying variables and equations on each of these model components. The object-oriented approach has the following phases:

1. *Define* the system briefly. What kind of system is it? What does it do?

2. *Decompose* the system into its most important *components*. Sketch model classes for these components, or use existing classes from appropriate libraries.

3. Define *communication*, that is, determine interactions and communication paths between these components.

4. Define *interfaces*, that is, determine the external ports/ *connectors* of each component for communication with other components. Formulate appropriate connector classes that allow a high degree of connectivity and reusability, while still allowing an appropriate degree of type checking of connections.

5. Recursively *decompose model components* of "high complexity" further into a set of "smaller" subcomponents by restarting from phase 2, until all components have been defined as instances of predefined types, library classes, or new classes to be defined by the user.

6. *Formulate new model classes* when needed, both base classes and derived classes:

   a. Declare new *model classes* for all model components that are not instances of existing classes. Each new class should include variables, equations, functions, and formulas relevant to define the behavior of the component, according to the principles already described for the first two phases of the traditional modeling approach.

   b. Declare possible *base classes* for increased reuse and maintainability by extracting common functionality and structural similarities from component classes with similar properties.

To get more of a feeling for how the object-oriented modeling approach works in practice, we will apply this method to modeling a simple tank system described in Section 4.2.

4.1.4 Top-Down Versus Bottom-Up Modeling

There are two related approaches to the structuring process when arriving at a model:

- *Top-down modeling.* This is useful when we know the application area rather well and have an existing set of library component models available. We start by defining the top-level system model, gradually decomposing into subsystems, until we arrive at subsystems that correspond to the component models in our library.

- *Bottom-up modeling.* This approach is typically used when the application is less well known, or when we do not have a ready-made library at hand. First, we formulate the basic equations and design small experimental models on the most important phenomena to gain a basic understanding of the application area. Typically, start with very simplified models and later add more phenomena. After some experimentation we have gained some understanding of the application area and can restructure our model fragments into a set of model components. These might have to be restructured several times if they turn out to have problems when used for applications. We gradually build more complex application models based on our components and finally arrive at the desired application model.

In the following we give examples of both top-down and bottom-up modeling. Section 4.3, on modeling of a DC motor from predefined components, gives a typical example of top-down modeling. The small tank modeling example described in Section 4.2 has some aspects of bottom-up modeling since we start with a simple flat tank model before creating component classes and forming the tank model. These examples gradually grow into a set of components that are used for the final application models.

4.1.5 Simplification of Models

It is sometimes the case that models are not precise enough to accurately describe the phenomena at hand. Too extensive approximations might have been done in certain parts of the model.

On the opposite side, even if we create a reasonable model according to the above methodology, it is not uncommon that parts of the model are too complex, which might lead to problems, for example:

- Too time-consuming simulations
- Numerical instabilities
- Difficulties in interpreting results due to too many low-level model details

Thus, there are usually good reasons to consider simplification of a model. It is sometimes hard to get the right balance between simplicity and preciseness of models. This is more an art than a science and requires substantial experience. The best way to acquire such experience, however, is to keep on designing models together with performing analysis and evaluation of their behavior. The following are some hints of where to look for model simplifications, for example, reduction of the number of state variables:

- *Neglect* small effects that are not important for the phenomena to be modeled.
- *Aggregate* state variables into fewer variables. For example, the temperatures at different points in a rod might sometimes be represented by the mean temperature of the whole rod.

Focus modeling on phenomena whose *time constants* are in the interesting range, that is:

- Approximate subsystems with very slow dynamics with constants.
- Approximate subsystems with very fast dynamics with static relationships, that is, not involving time derivatives of those rapidly changing state variables.

An advantage of ignoring the very fast and the very slow dynamics of a system is that the order of the model, that is, the number of state variables, is reduced. Systems having model components with time constants of the same order of magnitude are numerically simpler and more efficient to simulate. On the other hand, certain systems have the intrinsic property of having a great spread of time constants. Such systems give rise to stiff systems of differential equations, which require certain adaptive numerical solvers for simulation.

4.2 MODELING A TANK SYSTEM

Regarding our exercise in modeling methodology, let us consider a simple tank system example containing a level sensor and a controller that is controlling a valve via an actuator (Fig. 4.1). The liquid level h in the tank must be maintained at a fixed level as closely as possible. Liquid enters the tank through a pipe from a source, and leaves the tank via another pipe at a rate controlled by a valve.

4.2.1 Using the Traditional Approach

We first show the result of modeling the tank system using the traditional approach. A number of variables and equations related to the tank system have been collected into the flat model presented in the next section. At this stage we do not explain how the equations have been derived and what exactly certain variables mean. This is described in detail in subsequent sections where the object-oriented approach to modeling the tank system is presented.

4.2.1.1 Flat Tank System Model

The FlatTank model is a "flat" model of the tank system with no internal hierarchical "system structure" visible, that is, just a collection of variables and equations that model the system dynamics. The internal system structure consisting of components, interfaces, couplings between the components, and the like is not reflected by this model:

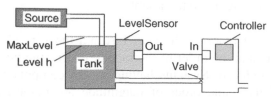

Figure 4.1 Tank system with a tank, a source for liquid, and a controller.

```
model FlatTank
   // Tank related variables and parameters
   parameter Real flowLevel(unit="m3/s")=0.02;
   parameter Real area(unit="m2")       =1;
   parameter Real flowGain(unit="m2/s") =0.05;
   Real          h(start=0,unit="m")    "Tank level";
   Real          qInflow(unit="m3/s")   "Flow through input
                                            valve";
   Real          qOutflow(unit="m3/s")  "Flow through output
                                            valve";
   // Controller related variables and parameters
   parameter Real K=2                "Gain";
   parameter Real T(unit="s")= 10        "Time constant";
   parameter Real minV=0, maxV=10;  //  Limits for flow output
   Real          ref=0.25  "Reference level for control";
   Real          error     "Deviation from reference level";
   Real          outCtr     "Control signal without limiter";
   Real          x;          "State variable for controller";
 equation
   assert (minV>=0,"minV must be greater or equal to zero");//
   der(h) = (qInflow-qOutflow)/area;  // Mass balance equation
   qInflow = if time>150 then 3*flowLevel else flowLevel;
   qOutflow = LimitValue(minV,maxV,-flowGain*outCtr);
   error = ref-h;
   der(x) = error/T;
   outCtr = K*(error+x);
 end  FlatTank;
```

A limiter function is needed in the model to reflect minimum and maximum flows through the output valve:

```
function LimitValue
   input    Real pMin;
   input    Real pMax;
   input    Real p;
   output   Real pLim;
   algorithm
     pLim := if p>pMax then pMax
              else if p<pMin then pMin
              else p;
end LimitValue;
```

Simulate the flat tank system model and plot the results (Fig. 4.2):

```
simulate(FlatTank, stopTime=250)
plot(h, stopTime=250)
```

4.2.2 Using the Object-Oriented Component-Based Approach

When using the object-oriented component-based approach to modeling, we first look for the internal structure of the tank system. Is

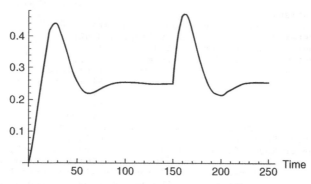

Figure 4.2 Simulation of the FlatTank model with plot of the tank level.

the tank system naturally decomposed into certain components? The answer is yes. Several components are actually visible in the structure diagram of the tank system in Figure 4.3, for example, the tank itself, the liquid source, the level sensor, the valve, and the controller.

Thus, it appears that we have six components. However, since we will choose very simple representations of the level sensor and the valve, that is, just a single scalar variable for each of these two components, we let these variables be simple Real variables in the tank model instead of creating two new classes containing a single variable each. Thus, we are left with three components that are to be modeled explicitly as instances of new classes: the tank, the source, and the controller.

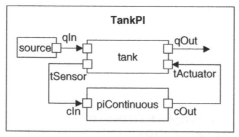

Figure 4.3 Tank system with a continuous PI controller and a source for liquid. Even though we have used arrows for clarity, there need not be a signal direction—there are only physical connections represented by equations.

The next stage is to determine the interactions and communication paths between these components. It is fairly obvious that fluid flows from the source to the tank via a pipe. Fluid also leaves the tank via an outlet controlled by the valve. The controller needs measurements of the fluid level from the sensor. Thus, a communication path from the sensor of the tank to the controller needs to be established.

Communication paths need to be connected somewhere. Therefore, connector instances need to be created for those components that are connected, and connector classes declared when needed. In fact, the system model should be designed so that the only communication between a component and the rest of the system is via connectors.

Finally, we should think about reuse and generalizations of certain components. Do we expect several variants of a component to be needed? In that case it is useful to collect the basic functionality into a base class, and let each variant be a specialization of that base class. In the case of the tank system we expect to plug in several variants of the controller, starting with a continuous proportional and integrating (PI) controller. Thus, it is useful for us to create a base class for tank system controllers.

4.2.3 Tank System with a Continuous PI Controller

The structure of the tank system model developed using the object-oriented component-based approach is clearly visible in Figure 4.3. The three components of the tank system—the tank, the PI controller, and the source of liquid—are explicit in Figure 4.3 and in the declaration of the class TankPI below.

```
model TankPI
   LiquidSource            source(flowLevel=0.02);
   PIcontinuousController  piContinuous(ref=0.25);
   Tank                    tank(area=1);
equation
   connect(source.qOut, tank.qIn);
   connect(tank.tActuator, piContinuous.cOut);
   connect(tank.tSensor, piContinuous.cIn);
end TankPI;
```

Tank instances are connected to controllers and liquid sources through their connectors. The tank has four connectors: qIn for input flow, qOut for output flow, tSensor for providing fluid level measurements, and tActuator for setting the position of the valve at the outlet of the tank. The central equation regulating the behavior of the tank is the *mass balance* equation, which in the current simple form assumes constant pressure. The output flow is related to the valve position by a flowGain parameter and by a limiter that guarantees that the flow does not exceed what corresponds to the open/closed positions of the valve.

```
model Tank
  ReadSignal    tSensor      "Connector, sensor reading tank level (m)";
  ActSignal     tActuator    "Connector, actuator controlling input flow";
  LiquidFlow    qIn          "Connector, flow (m3/s) through input valve";
  LiquidFlow    qOut         "Connector, flow (m3/s) through output valve";
  parameter Real area(unit="m2")      = 0.5;
  parameter Real flowGain(unit="m2/s") = 0.05;
  parameter Real minV=0, maxV=10; // Limits for output valve flow
  Real h(start=0.0, unit="m") "Tank level";
equation
  assert (minV>=0,"minV minimum Valve level must be >= 0");
  der(h)=(qIn.lflow-qOut.lflow)/area;// Mass balance equation
  qOut.lflow = LimitValue(minV,maxV,
    -flowGain*tActuator.act);
  tSensor.val = h;
end Tank;
```

As already stated, the tank has four connectors. These are instances of the following three connector classes:

```
connector ReadSignal "Reading fluid level"
  Real val(unit="m");
end ReadSignal;

connector ActSignal "Signal to actuator for setting valve position"
  Real act;
end ActSignal;

connector LiquidFlow "Liquid flow at inlets or outlets"
  Real lflow(unit="m3/s");
end LiquidFlow;
```

The fluid entering the tank must come from somewhere. Therefore, we have a liquid source component in the tank system. The flow increases sharply at time $= 150$ to factor of 3 of the previous flow level, which creates an interesting control problem that the controller of the tank has to handle.

```
model LiquidSource
  LiquidFlow qOut;
  parameter flowLevel = 0.02;
equation
  qOut.lflow = if time>150 then 3*flowLevel else flowLevel;
end LiquidSource;
```

The controller needs to be specified. We will initially choose a PI controller but later replace this by other kinds of controllers. The behavior of a continuous PI controller is primarily defined by the following two equations:

$$\frac{dx}{dt} = \frac{\text{error}}{T}$$
$$\text{outCtr} = K(\text{error} + x) \tag{4.1}$$

Here x is the controller state variable, error is the difference between the reference level and the actual level of obtained from the sensor, T is the time constant of the controller, outCtr is the control signal to the actuator for controlling the valve position, and K is the gain factor. These two equations are placed in the controller class `PIcontinuousController`, which extends the BaseController class defined later.

```
model PIcontinuousController
  extends BaseController(K=2,T=10);
  Real x "State variable of continuous PI controller";
equation
  der(x) = error/T;
  outCtr = K*(error+x);
end PIcontinuousController;
```

By integrating the first equation, thus obtaining x, and substituting into the second equation, we obtain the following expression for the control signal, containing terms that are directly proportional to the error signal and to the integral of the error signal, respectively, that is, a PI controller.

$$\text{outCtr} = K\left(\text{error} + \int \frac{\text{error}}{T} dt\right) \tag{4.2}$$

Both the PI controller and the proportional, integrating, derivative (PID) controller to be defined later inherit the partial controller class `BaseController`, containing common parameters, state variables, and two connectors: one to read the sensor and one to control the valve actuator.

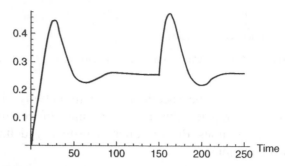

Figure 4.4   Tank level response for the `TankPI` system containing a `PI` controller.

In fact, the `BaseController` class also can be reused when defining discrete PI and PID controllers for the same tank example in Figure 4.3. Discrete controllers repeatedly sample the fluid level and produce a control signal that changes value at discrete points in time with a periodicity of Ts.

```
partial model BaseController
   parameter Real Ts(unit="s")=0.1   "Time period between discrete samples";
   parameter Real K=2                 "Gain";
   parameter Real T=10(unit="s")      "Time constant";
   ReadSignal     cIn                 "Input sensor level, connector";
   ActSignal      cOut                "Control to actuator, connector";
   parameter Real ref                 "Reference level";
   Real           error               "Deviation from reference level";
   Real           outCtr              "Output control signal";
equation
   error    = ref-cIn.val;
   cOut.act = outCtr;
end BaseController;
```

We simulate the `TankPI` model and obtain the same response as for the `FlatTank` model, which is not surprising given that both models have the same basic equations (Fig. 4.4):

```
simulate(TankPI, stopTime=250)
plot(tank.h)
```

4.2.4 Tank with Continuous PID Controller

We define a `TankPID` system as the same as the `TankPI` system except that the PI controller has been replaced by a PID controller. Here we see a clear advantage of the object-oriented component-based approach over the traditional modeling approach, since system

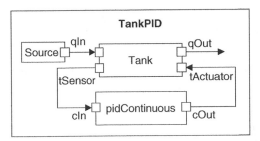

Figure 4.5   Tank system with the PI controller replaced by a PID controller.

components can be easily replaced and changed in a plug-and-play manner (Fig. 4.5).

The Modelica class declaration for the TankPID system appears as follows:

```
model TankPID
   LiquidSource               source(flowLevel=0.02);
   PIDcontinuousController  pidContinuous(ref=0.25);
   Tank                       tank(area=1);
equation
   connect(source.qOut, tank.qIn);
   connect(tank.tActuator, pidContinuous.cOut);
   connect(tank.tSensor, pidContinuous.cIn);
end TankPID;
```

A PID controller model can be derived in a similar way as was done for the PI controller. PID controllers react quicker to instantaneous changes than PI controllers due to the term containing the derivative. A PI controller on the other hand puts somewhat more emphasis on compensating slower changes. The basic equations for a PID controller are the following:

$$\frac{dx}{dt} = \frac{\text{error}}{T}$$
$$y = T\frac{d\,\text{error}}{dt} \tag{4.3}$$
$$\text{outCtr} = K\,(\text{error} + x + y)$$

We create a PIDcontinuousController class containing the three defining equations:

```
model PIDcontinuousController
   extends BaseController(K=2,T=10);
   Real x; // State variable of continuous PID controller
   Real y; // State variable of continuous PID controller
```

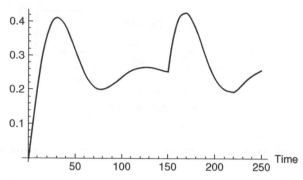

Figure 4.6   Tank level response for the `TankPID` system containing a PID controller.

```
equation
  der(x) = error/T;
  y      = T*der(error);
  outCtr = K*(error + x + y);
end  PIDcontinuousController;
```

By integrating the first equation and substituting x and y into the third equation, we obtain an expression for the control signal that contains terms directly proportional, proportional to the integral, and proportional to the derivative of the error signal, that is, a PID controller.

$$\text{outCtr} = K \left(\text{error} + \int \frac{\text{error}}{T} dt + T \frac{d\,\text{error}}{dt} \right) \qquad (4.4)$$

We simulate the tank model once more but now including the PID controller (Fig. 4.6):

```
simulate(TankPID, stopTime=250)
plot(tank.h)
```

The tank level as a function of time of the simulated system is quite similar to the corresponding system with a PI controller, but with somewhat faster control to restore the reference level after quick changes of the input flow.

The diagram in Figure 4.7 shows the simulation results from both `TankPI` and `TankPID` plotted in the same diagram for comparison.

```
simulate(compareControllers, stopTime=250)
plot({tankPI.h,tankPID.h})
```

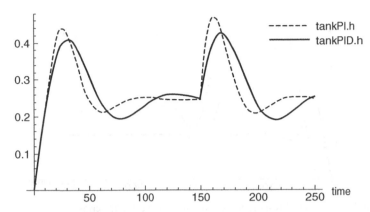

Figure 4.7 Comparison of `TankPI` and `TankPID` simulations.

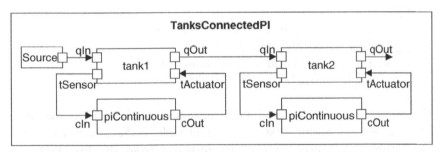

Figure 4.8 Two connected tanks with PI controllers and a source for liquid connected to the first tank.

4.2.5 Two Tanks Connected Together

The advantages of the object-oriented component-based approach to modeling become even more apparent when combining several components in different ways, as in the example depicted in Figure 4.8 where two tanks are connected in series, which is not uncommon in the process industry.

The Modelica model `TanksConnectedPI` corresponding to Figure 4.8 appears as follows:

```
model TanksConnectedPI
  LiquidSource   source(flowLevel=0.02);
  Tank           tank1(area=1);
  Tank           tank2(area=1.3);
  PIcontinuousController piContinuous1(ref=0.25);
  PIcontinuousController piContinuous2(ref=0.4);
```

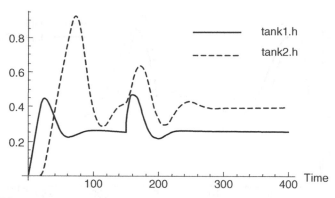

Figure 4.9 Tank level responses for two tanks connected in series.

```
equation
  connect(source.qOut,tank1.qIn);
  connect(tank1.tActuator,piContinuous1.cOut);
  connect(tank1.tSensor,piContinuous1.cIn);
  connect(tank1.qOut,tank2.qIn);
  connect(tank2.tActuator,piContinuous2.cOut);
  connect(tank2.tSensor,piContinuous2.cIn);
end TanksConnectedPI;
```

We simulate the connected tank system. We clearly see the tank level responses of the first and second tank to the changes in fluid flow from the source, where the response from the first tank, of course, appears earlier in time than the response of the second tank (Fig. 4.9):

```
simulate(TanksConnectedPI, stopTime=400)
plot({tank1.h,tank2.h})
```

4.3 TOP-DOWN MODELING OF A DC MOTOR FROM PREDEFINED COMPONENTS

In this section we illustrate the object-oriented component-based modeling process when using predefined library classes by sketching the design of a DC motor servo model. We do not go into any detail since the previous tank example was presented in quite some detail.

4.3.1 Defining the System

What does a DC motor servo do? It is a motor, the speed of which can be controlled by some kind of regulator (Fig. 4.10). Since it is a servo, it needs to maintain a specified rotational speed despite a varying load. Presumably, it contains an electric motor, some mechanical rotational transmissions and loads, some kind of control to regulate the rotational speed, and some electric circuits since the control system needs electric connections to the rest of the system, and there are electric parts of the motor. The reader may have noticed that it is hard to define a system without describing its parts, that is, we are already into the system decomposition phase.

4.3.2 Decomposing into Subsystems and Sketching Communication

In this phase we decompose the system into major subsystems and sketch communication between those subsystems. As already noted in the system definition phase, the system contains rotational mechanical parts including the motor and loads, an electric circuit model containing the electric parts of the DC motor together with its electric interface, and a controller subsystem that regulates the speed of the DC motor by controlling the current fed to the motor. Thus, there are three subsystems as depicted in Figure 4.11: controller, an electric circuit, and a rotational mechanics subsystem.

Figure 4.10   A DC motor servo.

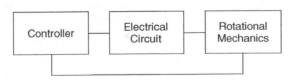

Figure 4.11   Subsystems and their connections.

Regarding the communication between the subsystems, the controller must be connected to the electric circuit since it controls the current to the motor. The electric circuit also needs to be connected to the rotational mechanical parts in order that electrical energy can be converted to rotational energy. Finally, a feedback loop including a sensor of the rotational speed is necessary for the controller to do its job properly. The commutation links are sketched in Figure 4.11.

4.3.3 Modeling the Subsystems

The next phase is to model the subsystems by further decomposition. We start by modeling the controller and manage to find classes in the standard Modelica library for a feedback summation node and a PI controller. We also add a step function block as an example of a control signal. All these parts are shown in Figure 4.12.

The second major component to decompose is the electric circuit part of the DC motor (Fig. 4.13). Here we have identified the standard parts of a DC motor such as a signal-controlled electric voltage generator, a ground component needed in all electric circuits, a resistor, an inductor representing the coil of the motor, and an electromotive force (emf) converter to convert electric energy to rotational movement.

The third subsystem, depicted in Figure 4.14, contains three mechanical rotational loads with inertia, one ideal gear, one rotational spring, and one speed sensor for measuring the rotational speed needed as information for the controller.

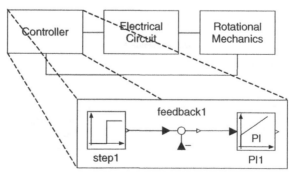

Figure 4.12   Modeling the controller.

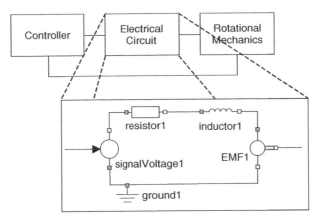

Figure 4.13 Modeling the electric circuit.

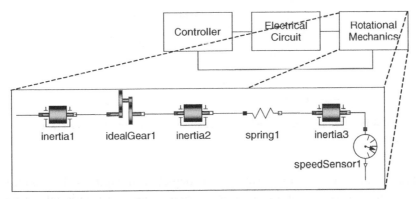

Figure 4.14 Modeling the mechanical subsystem including the speed sensor.

4.3.4 Modeling Parts in the Subsystems

We managed to find all the needed parts as predefined models in the Modelica class library. If that would not have been the case, we would also have needed to define appropriate model classes and identified the equations for those classes, as sketched for the parts of the control subsystem in Figure 4.15, the electric subsystem in Figure 4.16, and the rotational mechanics subsystem in Figure 4.17.

The electric subsystem depicted in Figure 4.16 contains electrical components with associated basic equations, for example, a resistor, an inductor, a signal voltage source, and an emf component.

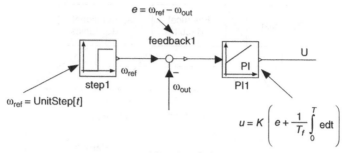

$$e = \omega_{ref} - \omega_{out}$$

feedback1

$$\omega_{ref}$$

step1

$$\omega_{ref} = \text{UnitStep}[t]$$

$$\omega_{out}$$

PI

U

PI1

$$u = K\left(e + \frac{1}{T_f}\int_0^T edt\right)$$

Figure 4.15   Basic equations and components in the control subsystem.

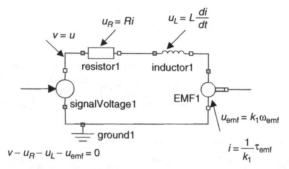

$$u_R = Ri$$

$$u_L = L\frac{di}{dt}$$

$$v = u$$

resistor1

inductor1

signalVoltage1

EMF1

ground1

$$u_{emf} = k_1\omega_{emf}$$

$$v - u_R - u_L - u_{emf} = 0$$

$$i = \frac{1}{k_1}\tau_{emf}$$

Figure 4.16   Defining classes and basic equations for components in the electric subsystem.

The rotational mechanics subsystem depicted in Figure 4.17 contains a number of components such as inertias, a gear, a rotational spring, and a speed sensor.

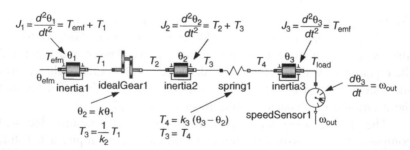

$$J_1 = \frac{d^2\theta_1}{dt^2} = T_{emf} + T_1$$

$$J_2 = \frac{d^2\theta_2}{dt^2} = T_2 + T_3$$

$$J_3 = \frac{d^2\theta_3}{dt^2} = T_{emf}$$

$$T_{efm}$$ $$\theta_1$$ $$T_1$$ $$T_2$$ $$\theta_2$$ $$T_3$$ $$T_4$$ $$\theta_3$$ $$T_{load}$$

$$\theta_{efm}$$

inertia1   idealGear1   inertia2   spring1   inertia3

$$\frac{d\theta_3}{dt} = \omega_{out}$$

speedSensor1 $$\omega_{out}$$

$$\theta_2 = k\theta_1$$

$$T_3 = \frac{1}{k_2}T_1$$

$$T_4 = k_3(\theta_3 - \theta_2)$$

$$T_3 = T_4$$

Figure 4.17   Defining classes and basic equations for components in the rotational mechanics subsystem.

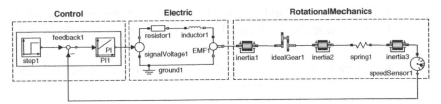

Figure 4.18 Finalizing the interfaces and connections between the subsystems, including the feedback loop.

4.3.5 Defining the Interfaces and Connections

When each subsystem has been defined from predefined models, the interfaces to the subsystem are given by the connectors of those components that interact with other subsystems. Each subsystem has to be defined to enable communication with the other subsystems according to the previously sketched commutation structure. This requires the connector classes to be carefully chosen to be type compatible. Actually, the selection and definition of these connector interfaces is one of the most important steps in the design of a model.

The completed model of the DC motor servo is depicted in Figure 4.18, including the three subsystems and the feedback loop.

4.4 DESIGNING INTERFACES–CONNECTOR CLASSES

As in all system design, defining the interfaces between the components of a system model is one of the most important design tasks since this lays the foundation for the communication between the components, thereby also heavily influencing the decomposition of the model.

Most important is to identify the basic *requirements* behind the design of component interfaces, that is, connector classes, that influence their structure. These requirements can be briefly stated as follows:

- It should be *easy* and *natural* to connect components. For interfaces to models of physical components it must be physically possible to connect those components.
- Component interfaces should facilitate *reuse* of existing model components in class libraries.

Fulfilling these goals requires careful design of connector classes. The number of connector classes should be kept small in order to avoid unnecessary mismatches between connectors due to different names and types of variables in the connectors.

Experience shows that it is surprisingly difficult to design connector classes that satisfy these requirements. There is a tendency for extraneous details that are created during software (model) development for various computational needs to creep into the interfaces, thereby making them harder to use and preventing component reuse. Therefore, one should keep the connector classes as simple as possible and try to avoid introducing variables that are not really necessary.

A good rule of thumb when designing connector classes for models of physical components is to identify the characteristics of (acausal) interaction in the physical real world between these components. The interaction characteristics should be simplified and abstracted to an appropriate level and reflected in the design of the connector classes. For nonphysical components, for example, signal blocks and software components in general, one has to work hard on finding the appropriate level of abstraction in the interfaces and trying these in practice for feedback on ease of use and reusability. The Modelica standard library contains a large number of well-designed connector classes that can serve as inspiration when designing new interfaces.

There are basically three different kinds of connector classes, reflecting three design situations:

1. If there is some kind of interaction between two *physical* components involving *energy flow*, a combination of one potential and one `flow` variable in the appropriate domain should be used for the connector class.
2. If information or *signals* are exchanged between components, `input/output` signal variables should be used in the connector class.

3. For complex interactions between components, involving several interactions of types 1 and 2 above, a hierarchically structured *composite connector* class is designed that includes one or more connectors of the appropriate type 1, 2, or 3.

When all the connectors of a component have been designed according to the three principles above, the formulation of the rest of the component class follows partly from the constraints implied by these connectors. However, these guidelines should not be followed blindly. There are several examples of domains with special conditions that deviate slightly from the above rules.

4.5 SUMMARY

In this chapter we have presented an object-oriented component-based process on how to arrive at mathematical models of the systems in which we are interested. System modeling is illustrated on a two-tank example, both using the flat approach and the object-oriented approach.

4.6 LITERATURE

General principles for object-oriented modeling and design are described in Rumbaugh (1991) and Booch (1991). A general discussion of block-oriented model design principles can be found in Ljung and Glad (1994), whereas Andersson (1994) describes object-oriented mathematical modeling in some depth.

Many concepts and terms in software engineering and in modeling/simulation are described in the standard software engineering glossary: (IEEE Std 610.12-1990) together with the IEEE standard glossary of modeling and simulation (IEEE Std 610.3-1989).

CHAPTER 5

The Modelica Standard Library

Much of the power of modeling with Modelica comes from the ease of reusing model classes. Related classes in particular areas are grouped into packages, which make them easier to find. This chapter gives a *quick overview* of some common Modelica packages.

A special package, called Modelica, is a standardized predefined package that together with the Modelica Language is developed and maintained by the Modelica Association. This package is also known as the *Modelica Standard Library*. It provides constants, types, connector classes, partial models, and model classes of components from various application areas, which are grouped into subpackages of the Modelica package.

The Modelica Standard Library can be used freely for both non-commercial and commercial purposes under the conditions of *The Modelica License* as stated in the front pages of this book. The full documentation as well as the source code of these libraries appear at the Modelica website: http://www.modelica.org/library/. These libraries are often included in Modelica tool distributions.

Version 3.1 of the Modelica standard library from August, 2009, contains about 920 models and blocks, and 610 functions in the sublibraries listed in Table 5.1.

A subpackage named Interfaces occurs as part of several packages. It contains *interface definitions* in terms of connector classes

Introduction to Modeling and Simulation of Technical and Physical Systems with Modelica,
First Edition. By Peter Fritzson
© 2011 the Institute of Electrical and Electronics Engineers, Inc. Published 2011 by John Wiley & Sons, Inc.

Table 5.1
Main Sublibraries of the Modelica Standard Library Version 3.1

Modelica.Electrical.Analog
Analog electrical and electronic components such as resistor, capacitor, transformers, diodes, transistors, transmission lines, switches, sources, sensors.

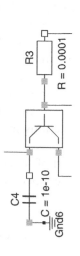

Modelica.Electrical.Digital
Digital electrical components based on VHDL (IEEE 1999) with nine valued logic. Contains delays, gates, sources, and converters between 2-, 3-, 4-, and 9-valued logic.

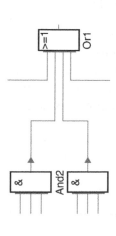

Modelica.Electrical.Machines
Uncontrolled, electrical machines, such as asynchronous, synchronous, and direct current motors and generators.

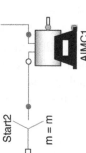

Modelica.Mechanics.Rotational

One-dimensional rotational mechanical systems, such as drive trains, planetary gear. Contains inertia, spring, gear box, bearing friction, clutch, brake, backlash, torque, etc.

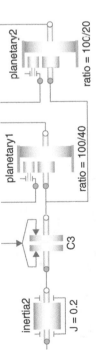

Modelica.Mechanics.Translational

One-dimensional translational mechanical systems, such as mass, stop, spring, backlash, and force.

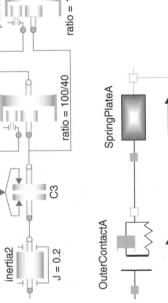

Modelica.Mechanics.MultiBody

Three-dimensional mechanical systems consisting of joints, bodies, force, and sensor elements. Joints can be driven by elements of the Rotational library. Every element has a default animation.

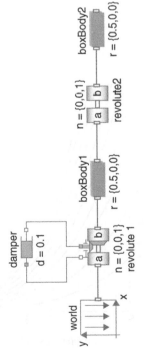

(continued)

Table 5.1
Main Sublibraries of the Modelica Standard Library Version 3.1 (*Continued*)

Modelica.Media

Large media library for single and multiple substance fluids with one and multiple phases:

- High-precision gas models based on the NASA Glenn coefficients + mixtures between these gas models
- Simple and high-precision water models (IAPWS/IF97)
- Dry and moist air models
- Table-based incompressible media.
- Simple liquid models with linear compressibility

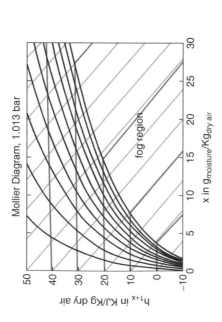

Mollier Diagram, 1.013 bar

h_{1+x} in KJ/Kg dry air

x in $g_{moisture}/Kg_{dry\ air}$

fog region

Modelica.Thermal

Simple thermo-fluid pipe flow, especially for machine cooling systems with water or air fluid. Contains pipes, pumps, valves, sensors, sources, etc. Furthermore, lumped heat transfer components are present, such as heat capacitor, thermal conductor, convection, body radiation, etc.

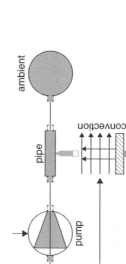

ambient

pipe

convection

pump

Modelica.Blocks
Continuous and discrete input/output blocks. Contains transfer functions, linear state space systems, non-linear, mathematical, logical, table, source blocks.

Modelica.StateGraph
Hierarchical state diagrams with similar modeling power as Statecharts. Modelica is used as synchronous "action" language. Deterministic behavior is guaranteed.

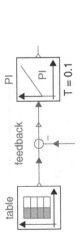

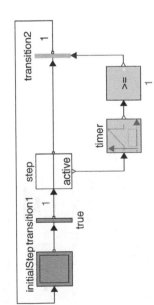

(continued)

Table 5.1
Main Sublibraries of the Modelica Standard Library Version 3.1 (*Continued*)

Modelica.Math.Matrices/Modelica.Utilities

Functions operating on matrices, e.g., to solve linear systems and compute eigen and singular values. Moreover, functions are provided to operate on strings, streams, and files.

```
import Modelica.Math.Matrices;

A = [1,2,3;
     3,4,5;
     2,1,4];
b = {10,22,12};
x = Matrices.solve(A,b);
Matrices.eigenValues(A);
```

Modelica.Constants, Modelica.Icons, Modelica.SIunits

Utility libraries to provide:

- Often used constants such as e, π, R.
- A library of icons that can be used in models
- About 450 predefined types, such as mass, angle, time, based on the international standard on units.

```
type Angle = Real (
  final quantity = "Angle",
  final unit    = "rad",
  displayUnit   = "deg");
```

for the application area covered by the package in question, as well as *commonly used partial classes* to be reused by the other subpackages within that package.

Also, a subpackage `Examples` contains *example models* on how to use the classes of the package in question in many libraries. Some libraries with sublibraries are shown in Table 5.2.

Table 5.2
Selected Libraries with Examples of Sublibraries `Interfaces` and `Examples`

`Modelica`	Standard library from the Modelica Association including the following sublibraries:

`Blocks`
`Interfaces`
`Continuous`
`. . .`
	Input/output blocks for use in block diagrams.
	Interfaces sublibrary to Blocks.
	Continuous control blocks with internal states.

`Electrical`
`Analog`
`Interfaces, Basic, Ideal,`
`Sensors, Sources,`
`Examples,`
`Lines, Semiconductors`
 	
`Digital`
`. . .`
	Common electrical component models
	Analog electrical component models
	Analog electrical sublibraries.
	Analog electrical sublibraries.
	Analog electrical sublibraries.
	Analog electrical sublibraries.
	
	Digital electrical components.

`Mechanics`
 	
`Rotational`
`Interfaces, Sensors,`
`Examples, . . .`
 	
`Translational`
`Interfaces, Sensors,`
`Examples`
 	
`. . .`
	General mechanical library.
	1D rotational mechanical component models.
	Rotational sublibraries.
	1D translational mechanical components.
	Translational sublibraries.
	
	3D mechanical systems—MultiBody library.
	MultiBody sublibraries.
	MultiBody sublibraries.
	MultiBody sublibraries.
	
	. . .

Table 5.3
Selection of Additional Free Modelica Libraries

`ModelicaAdditions:`	Old collection of additional Modelica libraries `Blocks`, `PetriNets, Tables, HeatFlow1D, MultiBody`.
`Blocks`	Additional input/output blocks.
`Discrete`	Discrete input/output blocks with fixed sample period.
`Logical`	Boolean input/output blocks.
`Multiplexer`	Combine and split-signal connectors of Real.
`PetriNets`	Petri nets and state transition diagrams.
`Tables`	Table lookup in 1 and 2 dimensions.
`HeatFlow1D`	1D heat flow.
`MultiBody`	3D mechanical systems—old MBS library with some connection restrictions and manual handling of kinematic loops.
`Interfaces,Joints,CutJoints`	Old MBS sublibraries
`Forces, Parts, Sensors,`	Old MBS sublibraries
`Examples`	Old MBS sublibrary
`SPOT`	Power systems in transient and steady-state mode, 2007
`ExtendedPetriNets`	Extended Petri net library, 2002
`ThermoFluid`	Old (superseded) library on thermodynamics and thermohydraulics, steam power plants, and process systems.
`SystemDynamics`	System dynamics a la J. Forrester, 2007.
`QSSFluidFlow`	Quasi-steady-state fluid flows.
`Fuzzy Control`	Fuzzy control library.
`VehicleDynamics`	Dynamics of vehicle chassis (obsolete), 2003 (replaced by commercial library)
`NeuralNetwork`	Neural network mathematical models, 2006
`WasteWater`	Wastewater treatment plants, 2003
`ATPlus`	Building simulation and control (fuzzy control included), 2005

Table 5.3
(*Continued*)

MotorCycleDynamics	Dynamics and control of motorcycles, 2009
SPICElib	Some capabilities of electric circuit simulator PSPICE, 2003
BondLib	Bond graph modeling of physical systems, 2007
MultiBondLib	Multibond graph modeling of physical systems, 2007
ModelicaDEVS	DEVS discrete event modeling, 2006
External.Media Library	External fluid property computation, 2008
VirtualLabBuilder	Implementation of virtual labs, 2007

There is also a number of free Modelica libraries that are not part of the Modelica Standard Library and have not yet been "certified" by the Modelica Association. The quality of these libraries is varying. Some are well documented and tested, whereas this is not the case for certain other libraries. The number of libraries available at the Modelica Association website is growing rapidly. Table 5.3 is a subset snapshot of the status in September 2009. Several commercial libraries, usually not free of charge, are also available.

When developing an application or a library in some application area, it is wise to use the standardized quantity types available in Modelica.SIunits and the standard connectors available in the corresponding Interfaces subpackage, for example, Modelica.Blocks.Interfaces of the Modelica Standard Library, in order that model components based on the same physical abstraction have compatible interfaces and can be connected together.

In Table 5.4 elementary connector classes are defined where potential variables are connector variables without the flow prefix and flow variables have the flow prefix.

In all domains two equivalent connectors are usually defined, for example, DigitalInput – DigitalOutput, HeatPort_a – HeatPort_b, and so forth. The variable declarations of these connectors are *identical*, only the icons are different in order that it is easy to distinguish two connectors of the same domain that are attached at same component model.

Table 5.4

Basic Connector Classes for Commonly Used Modelica Libraries

Domain	Potential Variables	Flow Variables	Connector Definition	Icons
Electrical analog	Electrical potential	Electrical current	Modelica.Electrical.Analog.Interfaces.Pin, .PositivePin, .NegativePin	
Electrical multiphase	Vector of electrical pins		Modelica.Electrical.MultiPhase.Interfaces.Plug, .PositivePlug, .NegativePlug	
Electrical space phasor	2 electrical potentials	2 electrical currents	Modelica.Electrical.Machines.Interfaces SpacePhasor	
Electrical digital	Integer (1.9)	—	Modelica.Electrical.Digital.Interfaces. DigitalSignal, .DigitalInput, DigitalOutput	
Translational	Distance	Cut-force	Modelica.Mechanics.Translational.Interfaces. Flange_a, .Flange_b	
Rotational	Angle	Cut-torque	Modelica.Mechanics.Rotational.Interfaces. Flange_a, .Flange_b	
3D mechanics	Position vector orientation object	Cut-force vector; cut-torque vector	Modelica.Mechanics.MultiBody.Interfaces. Frame, .Frame_a, .Frame_b, .Frame_resolve	

Domain	Symbols	Potential / Flow variables	Modelica interface
Simple fluid flow	◯ / ●	Pressure-specific enthalpy; mass flow rate, enthalpy flow rate	Modelica.Thermal.FluidHeatFlow.Interfaces. FlowPort, .FlowPort_a, .FlowPort_b
Heat transfer	□ / ■	Temperature; heat flow rate	Modelica.Thermal.HeatTransfer.Interfaces. HeatPort, .HeatPort_a, .HeatPort_b
Block diagram	△ △ △ / ▲ ▲ ▲	Real, Integer, Boolean	Modelica.Blocks.Interfaces.RealSignal, .RealInput, RealOutput.IntegerSignal, .IntegerInput, .IntegerOutput.BooleanSignal, .BooleanInput, .BooleanOutput
State machine	□ / ▲	Boolean variables (occupied, set, available, reset)	Modelica.StateGraph.Interfaces.Step_in, .Step_out, .Transition_in, .Transition_out
Thermofluid flow	◯ / ●	Pressure; Mass flow rate; Stream variables (if $m_{flow} < 0$): spec. enthalpy, mass fractions (m_i/m) extra property fractions (c_i/m)	Modelica_Fluid.Interfaces.FluidPort, .FluidPort_a, .FluidPort_b
Magnetic	□ / ■	Magnetic potential; Magnetic flux	Magnetic.Interfaces.MagneticPort, .PositiveMagneticPort,..NegativeMagneticPort

5.1 SUMMARY

This chapter has very briefly described the Modelica library structure, Version 3.1, as presented in the Modelica Language Specification 3.2 and at the Modelica Association web page at the time of this writing, including the Modelica Standard Library. The set of available libraries is growing quickly, however, the existing Modelica standard sublibraries are rather well-tested and have so far mostly gone through small evolutionary enhancements.

5.2 LITERATURE

All the free Modelica libraries described here, including both documentation and source code, can be found on the Modelica Association website, www.modelica.org. Documentation for several commercial libraries is also available on the Modelica website.

The most important reference for this chapter is Chapter 19 in the Modelica Language Specification (Modelica Association 2010), from which Tables 5.1 and 5.4 have been reused. Those tables were created by Martin Otter.

Glossary

algorithm section: part of a class definition consisting of the keyword `algorithm` followed by a sequence of statements. Like an equation, an algorithm section relates variables, i.e., constrains the values that these variables can take simultaneously. In contrast to an equation section, an algorithm section distinguishes inputs from outputs: An algorithm section specifies how to compute output variables as a function of given input variables. (See Section 2.14.)

array or **array variable:** variable that contains array elements. For an array, the ordering of its elements matters: The kth element in the sequence of elements of an array x is the array element with index k, denoted $x[k]$. All elements of an array have the same type. An array element may again be an array, i.e., arrays can be nested. An array element is hence referenced using n indices in general, where n is the number of dimensions of the array. Special cases are matrix ($n = 2$) and vector ($n = 1$). Array integer indices start with 1, not zero, i.e., the lower bound is 1. (See Section 2.13.)

array constructor: array can be built using the array function—with the shorthand curly braces $\{a, b, \ldots\}$, and can also include an iterator to build an array of expressions. (See Section 2.13.)

array element: element contained in an array. An array element has no identifier. Instead array elements are referenced by array access expressions called indices that use enumeration values or positive integer index values. (See Section 2.13.)

assignment: statement of the form x := expr. The expression expr must not have higher variability than x. (See Section 2.14.1.)

Introduction to Modeling and Simulation of Technical and Physical Systems with Modelica,
First Edition. By Peter Fritzson
© 2011 the Institute of Electrical and Electronics Engineers, Inc. Published 2011 by John Wiley & Sons, Inc.

attribute: property (or kind of record field) contained in a scalar variable, such as `min`, `max`, and `unit`. All attributes are predefined and attribute values can only be defined using a modification, such as in `Real x(unit="kg")`. Attributes cannot be accessed using dot notation and are not constrained by equations and algorithm sections. For example, in `Real x(unit="kg") = y;` only the values of x and y are declared to be equal, but not their unit attributes, nor any other attribute of x and y. (See Section 2.3.5.)

base class: class A is called a base class of B, if class B extends class A. This relation is specified by an extends clause in B or in one of B's base classes. A class inherits all elements from its base classes and may modify all nonfinal elements inherited from base classes. (See Section 3.7.)

binding equation: either a declaration equation or an element modification for the value of the variable. A variable with a binding equation has its value bound to some expression. (See Section 2.6.)

class: description that generates an object called instance. The description consists of a class definition, a modification environment that modifies the class definition, an optional list of dimension expressions if the class is an array class, and a lexically enclosing class for all classes. (See Section 2.3)

class type or **inheritance interface:** property of a class, consisting of a number of attributes and a set of public or protected elements consisting of element name, element type, and element attributes.

declaration assignment: assignment of the form `x := expression` defined by a variable declaration in a function. This is similar to a declaration equation, but different since an assigned variable usually can be assigned multiple times. (See Section 2.14.3.)

declaration equation: equation of the form `x = expression` defined by a component declaration. The expression must not have higher variability than the declared component x. Unlike other equations, a declaration equation can be overridden (replaced or removed) by an element modification. (See Section 2.6.)

derived class or **subclass:** class B is called derived from A, if B extends A. (See Section 2.4.)

element: part of a class definition, one of: class definition, component declaration, or extends clause. Component declarations and class definitions are called named elements. An element is either inherited from a base class or local.

element modification: part of a modification, overrides the declaration equation in the class used by the instance generated by the modified element. Example: vcc(unit="V")=1000. (See Sections 2.6 and 2.3.4.)

element redeclaration: part of a modification, replaces one of the named elements possibly used to build the instance generated by the element that contains the redeclaration. Example: redeclare type Voltage = Real(unit="V") replaces type Voltage. (See Section 2.5.)

encapsulated: class prefix that makes the class not depend on where it is placed in the package hierarchy, since its lookup is stopped at the class boundary. (See Section 2.16.)

equation: equality relation that is part of a class definition. A scalar equation relates scalar variables, i.e., constrains the values that these variables can take simultaneously. When $n - 1$ variables of an equation containing n variables are known, the value of the nth variable can be inferred (solved for). In contrast to a statement in an algorithm section, an equation does not define for which of its variable it is to be solved. Special cases are: initial equations, instantaneous equations, and declaration equations. (See Section 2.6)

event: something that occurs instantaneously at a specific time or when a specific condition occurs. Events are, for example, defined by the condition occurring in a when clause, if clause, or if expression. (See Section 2.15.)

expression: term built from operators, function references, variables/named constants, or variable references (referring to variables) and literals. Each expression has a type and a variability. (See Sections 2.1.1 and 2.1.3.)

extends clause: unnamed element of a class definition that uses a name and an optional modification to specify inheritance of a base class of the class defined using the class definition. (See Section 2.4.)

flattening: computation that creates a flattened class of a given class, where all inheritance, modification, etc. has been performed and all names resolved, consisting of a flat set of equations, algorithm sections, component declarations, and functions. (See Section 2.20.1.)

function: class of the specialized class function. (See Section 2.14.3.)

function subtype: class A is a function subtype of B iff A is a subtype of B and the additional formal parameters of function A that are not in function B are defined in such a way (e.g., additional formal parameters need to have default values) that A can be called at places where B is called. For more information, see Chapter 3 of Fritzson (2004) or Fritzson (2012).

identifier: atomic (not composed) name. Example: Resistor. (See Section 2.1.1 for names of variables.)

index or **subscript:** expression, typically of Integer type or the colon symbol (:), used to reference an element (or a range of elements) of an array. (See Section 2.13.)

inheritance interface or **class type:** property of a class, consisting of a number of attributes and a set of public or protected elements consisting of element name, element type, and element attributes. (See Sections 2.9 and 2.4.)

instance: object generated by a class. An instance contains zero or more components (i.e., instances), equations, algorithms, and local classes. An instance has a type. Basically, two instances have the same type, if their important attributes are the same and their public components and classes have pairwise equal identifiers and types. More specific type equivalence definitions are given, e.g., for functions. (See Section 2.13.)

instantaneous: equation or statement is instantaneous if it holds only at events, i.e., at single points in time. The equations and statements of a when clause are instantaneous. (See Section 2.15.)

literal: real, integer, Boolean, enumeration, or string constant value, i.e., a literal. Used to build expressions. (See Section 2.1.3.)

matrix: array where the number of dimensions is 2. (See Section 2.13.)

modification: part of an element. Modifies the instance generated by that element. A modification contains element modifications and element redeclarations. (See Section 2.3.4.)

name: Sequence of one or more identifiers. Used to reference a class or an instance. A class name is resolved in the scope of a class, which defines a set of visible classes. Example name: "Ele.Resistor". (See Sections 2.16 and 2.18.)

operator record: record with user-defined operations, defining, e.g., multiplication and addition. (See Section 2.14.4.)

partial: class that is incomplete and cannot be instantiated; useful, e.g., as a base class. (See Section 2.9.)

predefined type: one of the types Real, Boolean, Integer, String, and types defined as enumeration types. The attribute declarations of the predefined types define attributes such as min, max, and unit. (See Section 2.1.1.)

prefix: property of an element of a class definition that can be present or not be present, e.g., final, public, flow.

predefined type: one of the built-in types Real, Boolean, IntegerType, String, enumeration(...). (See Section 2.1.1.)

redeclaration: modifier with the keyword `redeclare` that changes a replaceable element. (See Section 2.5.)

replaceable: element that can be replaced by a different element having a compatible type. (See Section 2.5.)

restricted subtyping: type A is a restricted subtype of type B iff A is a subtype of B, and all public components present in A but not in B must be default connectable. This is used to avoid introducing, via a redeclaration, an unconnected connector in the object/class of type A at a level where a connection is not possible. (See Section 2.5.1 and Chapter 3 of Fritzson, 2004.)

scalar or **scalar variable:** variable that is not an array. (See Sections 2.1.1 and 2.13.)

simple type: `Real`, `Boolean`, `Integer`, `String`, and enumeration types. (See Section 2.1.1.)

specialized class: one of the following: model, connector, package, record, operator record, block, function, operator function, and type. The class specialization represents assertions regarding the content of the class and restricts its use in other classes, as well as providing enhancements compared to the basic class concept. For example, a class having the package class specialization must only contain classes and constants. (See Section 2.3.3.)

subtype compatible: relation between types. A is a subtype of B iff a number of properties of A and B are the same and all important elements of B have corresponding elements in A with the same names and their types being subtypes of corresponding element types in B. (See Chapter 3 of Fritzson, 2004.)

supertype: relation between types. The inverse of subtype. A is a subtype of B means that B is a supertype or base type of A. (See Section 2.4.)

type: property of an instance, expression, consisting of a number of attributes and a set of public elements consisting of element name, element type, and element attributes. Note: The concept of class type is a property of a class definition and also includes protected elements. Class type is used in certain subtype relationships, e.g., regarding inheritance. (See Chapter 3 of Fritzson, 2004.)

variability: property of an expression, which can have one of the following four values:

- *continuous*: an expression that may change its value at any point in time.
- *discrete*: may change its value only at events during simulation.

- *parameter*: constant during the entire simulation, but can be changed before each simulation and appears in tool menus. The default parameter values of models are often nonphysical and are recommended to be changed before simulation.
- *constant*: constant during the entire simulation; can be used and defined in a package.

Assignments `x:=expr` and binding equations `x=expr` must satisfy a variability constraint: The expression must not have a higher variability than variable x. (See Section 2.1.4.)

variable: instance (object) generated by a variable or constant declaration. Special kinds of variables are scalars, arrays, and attributes. (See Sections 2.1.1 and 2.3.1.)

variable declaration: element of a class definition that generates a variable, parameter, or constant. A variable declaration specifies (1) a variable name, i.e., an identifier; (2) the class to be flattened in order to generate the variable; and (3) an optional Boolean parameter expression. Generation of the variable is suppressed if this parameter expression evaluates to false. A variable declaration may be overridden by an element redeclaration. (See Sections 2.3.1, 2.1.1, and 2.1.3.)

variable reference: expression containing a sequence of identifiers and indices. A variable reference is equivalent to the referenced object. A variable reference is resolved (evaluated) in the scope of a class (or expression for the case of a local iterator variable). A scope defines a set of visible variables and classes. Example reference: `Ele.Resistor.u[21].r` (See Sections 2.1.1 and 2.13.)

vector: array where the number of dimensions is 1. (See Section 2.13.)

LITERATURE

This glossary has been slightly adapted from the Modelica language specification 3.2 (Modelica Association 2010). The first version of the glossary was developed by Jakob Mauss. The current version contains contributions from many Modelica Association members.

OpenModelica and OMNotebook Commands

This appendix gives a short overview of the OpenModelica commands, and a quick introduction to the OMNotebook electronic book, which can be used for Modelica textual modeling.

B.1 OMNOTEBOOK INTERACTIVE ELECTRONIC BOOK

Interactive electronic notebooks are active documents that may contain technical computations and text as well as graphics. Hence, these documents are suitable to be used for teaching and experimentation, simulation scripting, model documentation and storage, and the like. OMNotebook is an open-source implementation of such an electronic book, which belongs to the OpenModelica tool set.

- OMNotebook and the DrModelica documents are automatically installed when you install OpenModelica. To start OMNotebook on Windows use the program menu `OpenModelica->OMNotebook`, or double click on the `.onb`

Introduction to Modeling and Simulation of Technical and Physical Systems with Modelica,
First Edition. By Peter Fritzson
© 2011 the Institute of Electrical and Electronics Engineers, Inc. Published 2011 by John Wiley & Sons, Inc.

175

file you would like to open. The `DrModelica.onb` document is automatically opened when you start OMNotebook.

- To evaluate a cell, just click in the specific cell and press Shift + Enter. You can also evaluate a sequence of cells by clicking on a cell marker to the right, and pressing Shift + Enter.

- If you end a command by a semicolon (;), the value of the command will not be returned in an output cell.

- When using or saving your own files it is useful to first change the directory to the path where your files are located. This can be done by the `cd()` command.

- To perform a simulation, first evaluate the cell (cells) containing the model by clicking on it and pressing Shift + Enter. Then you have to evaluate a simulate command, for example, by typing `"simulate(modelname, startTime=0, stopTime=25 );` in a Modelica input cell, and pressing Shift + Enter.

- You can save typing by just writing the initial part of a command, e.g., sim for simulate, and push shift-tab. Then the command will be automatically expanded and completed.

- When writing Modelica code, a special `ModelicaInput` cell must be used.

- You can create new input cells by the `Cell->Input Cell` pull-down menu command, or by the short-cut command ctrl-shift-I, whereas a new cell with the same text style as the one above can be created by the short-cut command Alt + Enter.

After simulating a class, it is possible to plot or just look at the values of the variables in the class by evaluating the plot command or the val command.

Variable names given to the plot command refer to the most recently simulated model—you do not need to provide the modelname as a prefix.

For a more extensive tutorial explanation on how to use a notebook, see the notebook chapter in the OpenModelica Users Guide, which is included in the OpenModelica installation and can be reached under the (Windows) program menu item for OpenModelica.

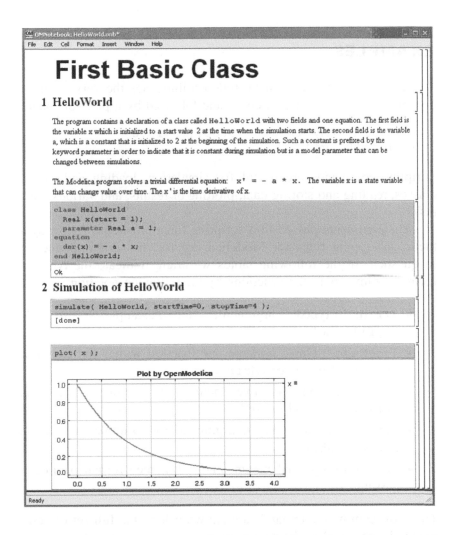

The DrModelica notebook has been developed to facilitate learning the Modelica language as well as providing an introduction to object-oriented modeling and simulation. It is based on, and is supplementary material to, the Modelica book: *Principles of Object-Oriented Modeling and Simulation with Modelica* (Fritzson 2004). All of the examples and exercises in DrModelica and the page references are from that book. Most of the text in DrModelica is also based on that book.

B.2 COMMON COMMANDS AND SMALL EXAMPLES

The following session in OpenModelica illustrates the very common combination of a simulation command followed by a plot command:

```
> simulate(myCircuit, stopTime=10.0)
> plot({R1.v})
```

Scripting commands are useful for simulation, loading and saving classes, reading and storing data, plotting of results, and various other tasks.

The arguments passed to a scripting function should follow syntactic and typing rules for Modelica and for the scripting function in question. In the following tables we briefly indicate the types of formal parameters to the functions by the following notation:

- `String` typed argument, for example, `"hello"`, `"myfile.mo"`
- `TypeName`—class, package, or function name, for example, `MyClass`, `Modelica.Math`
- `VariableName`—variable name, for example, `v1`, `v2`, `vars1[2].x`, and so forth
- `Integer` or `Real` typed argument, for example, `35`, `3.14`, `xintvariable`
- `options`—optional parameters with named formal parameter passing

The most common commands are shown below; the full set of commands is presented in the next section.

`simulate` (*className, options*). Translate and simulate a model, with optional start time, stop time, and optional number of simulation intervals or steps for which the simulation results will be computed. Many steps will give higher time resolution, but occupy more space and take longer to compute. The default number of intervals is 500.

Inputs: TypeName *className*; Real *startTime*;
Real *stopTime*; Integer *numberOfIntervals*;
Real *outputInterval*; String *method*;

Real *tolerance*; Real *fixedStepSize*;
Outputs: SimulationResult simRes;
Example 1: simulate(myClass);
Example 2: simulate(myClass, startTime=0,
stopTime=2, numberOfIntervals=1000,
tolerance=1e-10);.

plot(*variables*, options). Plot a single variable or several variables from the most recently simulated model, where *variables* is a single name or a vector of variable names if several variables should be plotted with one curve each. The optional parameters xrange and yrange allows you to specify the plot intervals in the diagram.

Inputs: VariableName *variables*; String *title*;
Boolean *legend*; Boolean *gridLines*;
Real *xrange*[2] i.e. {xmin,xmax};
Real *yrange*[2] i.e. {ymin,ymax};
Outputs: Boolean res;
Example 1: plot(x)
Example 2: plot(x,xrange={1,2},yrange={0,10})
Example 3: plot({x,y,z})   // Plot 3 curves, for x, y, and z.

quit(). Leave and quit the OpenModelica environment.

B.3 COMPLETE LIST OF COMMANDS

Below is the complete list of basic OpenModelica scripting commands. An additional set of more advanced commands for use by software clients is described in the OpenModelica System Documentation.

First, we show an interactive OpenModelica session using a few of the commands. The interactive command line interface can be used from the OpenModelica tools OMNotebook, OMShell, or from the command line in the OpenModelica MDT Eclipse plug-in.

```
>> model Test Real y=1.5; end Test;
{Test}

>> instantiateModel(Test)
"fclass Test
Real y = 1.5;
end Test;
"
```

```
>> list(Test)
  "model Test
     Real y=1.5;
  end Test;
  "

>> plot(y)

>> a:=1:10
  {1,2,3,4,5,6,7,8,9,10}

>> a*2
  {2,4,6,8,10,12,14,16,18,20}

>> clearVariables()
  true

>> clear()
  true

>> getClassNames()
  {}
```

Here is the complete list of basic commands in OpenModelica.

cd() Return the current directory as a string.
Outputs: String *dir*;
cd(dir) Change directory to directory *dir* given as a string.
Inputs: String *dir*;
Outputs: Boolean *res*;
Example: cd("C:\MyModelica\Mydir")
checkModel (className) Flatten model, optimize equations, and report errors.
Input: TypeName *className*;
Outputs: Boolean *res*;
Example: checkModel(myClass)
clear() Clears all loaded definitions, including variables and classes.
Outputs: Boolean *res*;
clearVariables() Clear all user-defined variables.
Outputs: Boolean *res*;
clearClasses() Clear all class definitions.
Outputs: Boolean *res*;
clearLog() Clear the log.
Outputs: Boolean res;

closePlots() Close all plot windows.

Outputs: `Boolean res;`

dumpXMLDAE (`modelname`,...) Export an XML representation of a flattened and optimized model, according to several optional parameters.

exportDAEtoMatlab (`name`) Export a Matlab representation of a model.

getLog() Return log as a string.

Outputs: `String log;`

help() Print help text of commands returned as a string.

instantiateModel (`modelName`) Flatten model, and return a string containing the flat class definition.

Input: `TypeName className;`

Outputs: `String flatModel;`

list() Return a string containing all loaded class definitions.

Outputs: `String classDefs;`

list(`className`) Return a string containing the class definition of the named class.

Input: `TypeName className;`

Outputs: `String classDef;`

listVariables() Return a vector of the names of the currently defined variables.

Outputs: `VariableNam[:] names;`

Example: `listVariables()    returns {x,y, ...}`

loadFile (`fileName`) Load Modelica file (.mo) with name given as string argument *filename*.

Input: `String fileName` *Outputs*: `Boolean res;`

Example: `loadFile("../myLibrary/myModels.mo"`

loadModel (`className`) Load the file corresponding to the *className*, using the Modelica class name to filename mapping to locate the file, searching from the path indicated by the environment variable OPENMODELICALIBRARY

Note: if, e.g., `loadModel(Modelica)` fails, you may have OPENMODELICALIBRARY pointing at the wrong location.

Input: `TypeName className`

Outputs: `Boolean res;`

Example1: `loadModel(Modelica.Electrical)`

plot(`variables`, `options`) Plot a single variable or several variables from the most recently simulated model, where *variables* is single name or a vector of variable names if several variables should be plotted, one curve each. The optional parameters `xrange` and `yrange` allows you to specify the plot intervals in the diagram.

Input: VariableName *variables*; String *title*;
   Boolean *legend*; Boolean *gridLines*;
   Real *xrange*[2], i.e., {xmin,xmax};
   Real *yrange*[2], i.e., {ymin,ymax};
Outputs: Boolean res;
Example 1: plot(x)
Example 2: plot(x,xrange={1,2},yrange={0,10})
Example 3: plot({x,y,z})   // Plot 3 curves, for x, y, and z.

plotParametric (*variables1*, *variables2*, options)
Plot each pair of corresponding variables from the vectors of variables or single variables *variables1*, *variables2* as a parametric plot.

Input: VariableName *variables1*[:];
VariableName
variables2[size(variables1,1)]; String *title*;
Boolean *legend*; Boolean *gridLines*;
Real *range*[2,2];
Outputs: Boolean *res*;
Example 1: plotParametric(x,y)
Example 2: plotParametric({x1,x2,x3}, {y1,y2,y3})

plot2(*variables*, options) Another implementation (in Java) called plot2, that supports most of the options of plot().

plotParametric2 (*variables1*, *variables2*, options)
Another function implementation (in Java) called plotParametric2, that supports most of the options of plotParametric.

plotVectors(v1, v2, options) Plot vectors v1 and v2 as an x-y plot. *Inputs*: VariableName v1;
VariableName v2; *Outputs*: Boolean res;

quit() Leave and quit the OpenModelica environment.

readFile (*fileName*) Load file given as string *fileName* and return a string containing the file contents.

Input: String fileName; String matrixName;
int nRows; int nColumns;
Outputs: Real res[nRows,nColumns];
Example 1: readFile("myModel/myModelr.mo")

readMatrix (*fileName*, *matrixName*) Read a matrix from a file given *fileName* and *matrixName*.

Input: String *fileName*; String *matrixName*;
Outputs: Boolean matrix[:,:];
readMatrix (fileName, matrixName, nRows, nColumns)
Read a matrix from a file, given filename, matrix name, #rows, and #columns.
Input: String fileName; String matrixName;
int nRows; int nColumns;
Outputs: Real res[nRows,nColumns];
readMatrixSize (*fileName*, *matrixName*) Read the matrix dimension from a *file* given a *matrix name*.
Input: String *fileName*; String *matrixName*;
Outputs: Integer sizes[2];
readSimulation Result(*fileName*, *variables*, *size*)
Read the simulation result for a list of variables and return a matrix of values (each column as a vector or values for a variable.) The size of the result (previously obtained from calling readSimulation-ResultSize is given as input.
Input: String *fileName*; VariableName *variables*[:];
Integer *size*;
Outputs: Real res[size(variables,1),size)];
readSimulation ResultSize (*fileName*) Read the size of a simulation result, i.e., the number of computed and stored simulation points of the trajectory vector, from a file.
Input: String fileName; *Outputs*: Integer size;
runScript (*fileName*) Execute script file with filename given as string argument *fileName*.
Input: String fileName; *Outputs*: Boolean res;
Outputs: runScript("simulation.mos")
saveLog(*fileName*) Save the simulation log with error messages to a file.
Input: String fileName; *Outputs*: Boolean res;
saveModel (*fileName*, *className*) Save the model/class with name *className* in the file given by the string argument *fileName*.
Input: String *fileName*; TypeName *className*
Outputs: Boolean *res*;
saveTotalModel (*fileName*, *className*) Save total class definition into file of a class.
Input: String *fileName*; TypeName *className*;
Outputs: Boolean res;

simulate (*className, options*) Translate and simulate a model, with optional start time, stop time, and optional number of simulation intervals or steps for which the simulation results will be computed. Many steps will give higher time resolution, but occupy more space and take longer to compute. The default number of intervals is 500.

Input: TypeName *className*; Real *startTime*;
Real *stopTime*; Integer *numberOfIntervals*;
Real *outputInterval*; String *method*;
Real *tolerance*; Real *fixedStepSize*;
Outputs: SimulationResult simRes;
Example 1: simulate(myClass);
Example 2: simulate(myClass, startTime=0,
stopTime=2, numberOfIntervals=1000,
tolerance=1e-10);.

system(*str*) Execute *str* as a system(shell) command in the operating system; return integer success value. Output into stdout from a shell command is put into the console window.

Input: String *str*; *Outputs*: Integer res;
Example: system("touch myFile")

timing(*expr*) Evaluate expression *expr* and return the number of seconds (elapsed time) of the evaluation.

Input: Expression *expr*; *Outputs*: Integer res;
Example: timing(x*4711+5)

translateModel (*className*) Flatten model, optimize equations, and generate code.

Input: TypeName *className*;
Outputs: SimulationObject res;

typeOf(*variable*) Return the type of the *variable* as a string.

Input: VariableName *variable*;
Outputs: String res;
Example: typeOf(R1.v)

val(*variable, timepoint*) Return the value of the simulation result *variable* evaluated or interpolated at the *timepoint*. Results from the most recent simulation are used.

Input: VariableName *variable*; Real *timepoint*;
Outputs: Real res;
Example 1: val(x,0)
Example 2: val(y.field,1.5)

writeMatrix (*fileName*, *matrixName*, *matrix*)   Write
matrix to file given a matrix name and a matrix.
  Input: String *fileName*;
    String *matrixName*; Real matrix[:,:];
  Outputs: Boolean *res*;

B.4 OMSHELL AND DYMOLA

OMShell

OMShell is a very simple command line interface to OpenModelica.
Note that OMNotebook is usually recommended for beginners since
it has more error checking. OMShell has the following extra facilities
to navigate among commands:

- Exit OMShell by pressing Ctrl-d.
- Up arrow—Get previously given command.
- Down arrow—Get next command.
- Tab—Command completion of the builtin OpenModelica com-
  mands.
- Circulate through the commands by only using tab key.

Dymola Scripting

Dymola is a widely used commercial modeling and simulation tool for
Modelica. The scripting language for Dymola is similar to the Open-
Modelica one, but there are some differences, most notably the use
of strings, for example, "Modelica.Mechanics", for class names
and variable names instead of using the names directly, for example,
Modelica.Mechanics, as in OpenModelica scripting.
  Below is an example of a Dymola script file for the
CoupledClutches example in the standard Modelica library. For
a complete list of Dymola scripting commands, consult the Dymola
users guide.

```
translateModel("Modelica.Mechanics.Rotational.
  Examples.CoupledClutches")
experiment(StopTime=1.2)
simulate
plot({"J1.w","J2.w","J3.w","J4.w"});
```

LITERATURE

An overview of OpenModelica can be found in Fritzson et al. (2005). Literate Programming (Knuth 1984) is a form of programming where programs are integrated with documentation in the same document. Mathematica notebooks (Wolfram 1997) is one of the first WYSIWYG (What-You-See-Is-What-You-Get) systems that support Literate Programming. Such notebooks were used early with Modelica, for example, in the MathModelica modeling and simulation environment; see Fritzson (2006) and Chapter 19 in Fritzson (2004). The DrModelica notebook has been developed to facilitate learning the Modelica language as well as providing an introduction to object-oriented modeling and simulation. It is based on, and is supplementary material to, the Modelica book (Fritzson 2004). Dymola (Dassault Systemes 2011) is an industrial-strength tool for modeling and simulation. MathModelica (MathCore 2011) is another commercial tool for Modelica modeling and simulation.)

Textual Modeling with OMNotebook and DrModelica

This appendix presents a few textual modeling exercises with Modelica that can, for example, be used in a minicourse on modeling and simulation. It is particularly simple to run the exercises in the OMNotebook electronic book, which is part of OpenModelica, downloadable from www.openmodelica.org. After installation, start OMNotebook from the menu OpenModelica->OMNotebook. You can find OMNotebook and OpenModelica commands in Appendix B.

When OMNotebook is started, the DrModelica notebook is automatically opened. This notebook has been developed to facilitate learning the Modelica language as well as providing an introduction to object-oriented modeling and simulation. It is based on, and is supplementary material to, the Modelica book (Fritzson 2004). All of the examples and exercises in DrModelica and the page references are from that book. Most of the text in DrModelica is also based on that book.

The following set of exercises, downloadable from the web page of this book at www.openmodelica.org, can be used with any Modelica tool. If OpenModelica is used, they can be accessed by opening the document TextualModelingExercises.onb in the testmodels

Introduction to Modeling and Simulation of Technical and Physical Systems with Modelica,
First Edition. By Peter Fritzson
© 2011 the Institute of Electrical and Electronics Engineers, Inc. Published 2011 by John Wiley & Sons, Inc.

directory in the OpenModelica installation, for example, by double-clicking on the file or using the `File->Open` pull-down menu command in OMNotebook.

C.1 HELLOWORLD

Simulate and plot the following example in the exercise notebook with one differential equation and one initial condition. Do a slight change in the model, resimulate, and replot.

```
model HelloWorld   "A simple equation"
   Real x(start=1);
equation
   der(x)= -x;
end HelloWorld;
```

Give a partial simulate command, for example, `simul`, in an input cell (can be created by ctrl-shift-I) in the notebook, push shift-tab for command completion, fill in the name HelloWorld, and simulate it!

Before command completion:

```
simul
```

After command completion using shift-tab:

```
simulate(modelname, startTime=0, stopTime=1,
numberOfIntervals=500, tolerance=1e-4)
```

After filling in HelloWorld:

```
simulate(HelloWorld, startTime=0, stopTime=1,
numberOfIntervals=500, tolerance=1e-4)
```

Fill in a plot command in an input cell (can also be expanded by command completion):

```
plot(x)
```

Take a look at the interpolated value of the variable x at `time=0.5` using the `val(variableName, time)` function:

```
val(x,0.5)
```

Also take a look at the value at `time=0.0`:

```
val(x,0.0)
```

C.2 TRY DRMODELICA WITH VANDERPOL AND DAEEXAMPLE MODELS

Locate the `VanDerPol` model in DrModelica (link from DrModelica Section 2.1), run it, change it slightly, and rerun it.

Change the simulation `stopTime` to 10, then simulate and plot.

Change the `lambda` parameter in the model to 10, then simulate for 50 s and plot. Why is the plot looking like this?.

Locate the DAEExample in DrModelica. Simulate and plot.

C.3 SIMPLE EQUATION SYSTEM

Make a Modelica model that solves the following equation system with initial conditions, simulate, and plot the results:

$$\dot{x} = 2*x*y - 3*x$$
$$\dot{y} = 5*y - 7*x*y$$
$$x(0) = 2$$
$$y(0) = 3$$

C.4 HYBRID MODELING WITH BOUNCINGBALL

Locate the BouncingBall model in one of the hybrid modeling sections of DrModelica (e.g., the when equations link in Section 2.9), run it, plot the curves, change it slightly, rerun it, plot it again, and observe the difference.

Locate the BouncingBall model in one of the hybrid modeling sections of DrModelica (the when equations link in Section 2.9), run it, change it slightly, and rerun it.

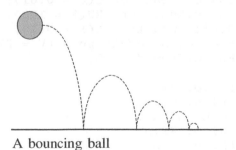

A bouncing ball

C.5 HYBRID MODELING WITH SAMPLE

Make a square signal with a period of 1 s and that starts at $t = 2.5$ s. Note that it is possible to use either an equation or an algorithm solution. Hint: An easy way is to use sample(...) to generate events, and define a variable that switches sign at each event.

C.6 FUNCTIONS AND ALGORITHM SECTIONS

1. Write a function, sum, which calculates the sum of real numbers, for a vector of arbitrary size.

2. Write a function, average, which calculates the average of real numbers, in a vector of arbitrary size. The function average should make use of a function call to sum.

C.7 ADDING A CONNECTED COMPONENT TO AN EXISTING CIRCUIT

Add a capacitor between the R2 component and the R1 component and an inductor between the R1 and voltage component. Use the SimpleCircuit model below and Modelica standard library components.

```
loadModel(Modelica);

model SimpleCircuit
  import Modelica.Electrical.Analog;
  Analog.Basic.Resistor   R1(R = 10);
  Analog.Basic.Capacitor  C(C = 0.01);
  Analog.Basic.Resistor   R2(R = 100);
  Analog.Basic.Inductor   L(L = 0.1);
  Analog.Sources.SineVoltage AC(V = 220);
  Analog.Basic.Ground     G;
equation
  connect(AC.p, R1.p);
  connect(R1.n, C.p);
  connect(C.n, AC.n);
  connect(R1.p, R2.p);
  connect(R2.n, L.p);
```

```
connect(L.n, C.n);
connect(AC.n, G.p);
end SimpleCircuit;
```

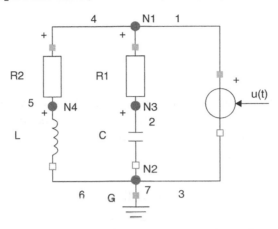

This example illustrates that it can sometimes be rather inconvenient to use textual modeling compared to graphical modeling. If you wish, you can verify the result using a graphical model editor as described in Appendix D.

C.8 DETAILED MODELING OF AN ELECTRIC CIRCUIT

This exercise consists of building a number of electrical components. Here follows the equations that describe each component. You can skip the equations subsection if you are already familiar with the equations.

C.8.1 Equations

The *ground element*

$$v_p = 0$$

where v_p is the potential of the ground element.

A *resistor*

$$i_p + i_n = 0$$

$$u = v_p - v_n$$

$$u = Ri_p$$

where i_p and i_n represent the currents into the positive and negative pin (or port) of the resistor, v_p and v_n the corresponding potentials, u the voltage over the resistor, and R the resistance.

An *inductor*

$$i_p + i_n = 0$$
$$u = v_p - v_n$$
$$u = Li_p'$$

where i_p and i_n represent the currents into the positive and negative pin (or port) of the inductor, v_p and v_n the corresponding potentials, u the voltage over the inductor, L the inductance, and i_p' the derivative of the positive pin current.

A *voltage source*

$$i_p + i_n = 0$$
$$u = v_p - v_n$$
$$u = V$$

where i_p and i_n represent the currents into the positive and negative pin (or port) of the voltage source, v_p and v_n the corresponding potentials, u the voltage over the voltage source, and V the constant voltage.

C.8.2 Implementation

Build Modelica models for the above-mentioned model components (ground element, resistor, inductor, voltage source, etc.). A connector class representing an electrical pin should ri4w5 be defined. Observe that the first two equations defining each electrical component with two pins above are equal. Utilize this observation to define a partial model, TwoPin, to be used in the definition of any electrical two-pin component. Hence a total of six components (Pin, Ground, TwoPin, Resistor, Inductor, and a VoltageSource) should be built.

Use the defined components to build a model of a circuit diagram and simulate the behavior of the circuit.

User-Defined Types

First define the types Voltage and Current:

```
type Voltage = Real;
type Current = Real;
```

Pin

The Pin has a potential, v, and current variable, i. According to Kirchhoff's laws, potentials are set equal and currents summed to zero at connections. Hence, v is a potential nonflow variable and i is a flow variable:

```
connector Pin
  ...
  ...
end Pin;
```

Ground

The Ground component has a positive Pin and a simple equation:

```
model Ground
  Pin p;
equation
  ...
end Ground;
```

TwoPin

The TwoPin element has a positive and negative Pin, a voltage u and a current i defined (the current i does not appear in the equations above and is only introduced to simplify notation):

```
model TwoPin
  Pin p, n;
  ...
  ...
equation
  ...
  ...
  ...
end TwoPin;
```

Resistor

To define the resistor, the partial model TwoPin is extended, and we only add a declaration of the parameter R together with Ohm's law that relates voltage and current to each other:

```
model Resistor
  extends TwoPin;
```

```
    . . .
equation
    . . .
end Resistor;
```

An equivalent model without use of a partial model would appear as follows:

```
model Resistor
    . . .
    . . .
    . . .
    . . .
    . . .
equation
    . . .
    . . .
    . . .
    . . .
end Resistor;
```

Note: The extends clause in the Modelica language can be thought of as just copying and pasting information from the partial model.

Inductor

The equation relating voltage and current for an inductor together with the inductance L are added to the partial model:

```
model Inductor
    . . .
    . . .
equation
    . . .
end Inductor;
```

VoltageSource

Here the partial model is extended with the trivial equation that the voltage between the positive and negative pins of the voltage source is kept constant:

```
model VoltageSource
    . . .
    . . .
equation
```

```
  . . .
end VoltageSource;
```

C.8.3 Putting the Circuit Together

Below is an example of a simple circuit where we instantiate the parameters of the components to other values than the default:

```
model Circuit
  Resistor R1(R=0.9);
  Inductor L1(L=0.01);
  Ground G;
  VoltageSource EE(V=5);
equation
  connect(EE.p, R1.p);
  connect(R1.n, L1.p);
  connect(L1.n, G.p);
  connect(EE.n, G.p);
end Circuit;
```

C.8.4 Simulation of the Circuit

Simulate the circuit:

```
simulate(Circuit, startTime=0, stopTime=1)
```

Several signals can be plotted, for example, R1.i, which is the current through the resistor R1:

```
plot(R1.i)
```

Graphical Modeling Exercises

The following small graphical modeling examples can be used with any Modelica tool graphic model editor.

D.1 SIMPLE DC MOTOR

Make a simple DC motor using the Modelica standard library that has the following structure:

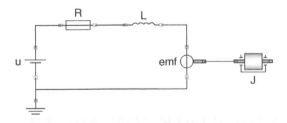

Save the model from the graphic editor, either simulate it directly from the graphic editor or load it and simulate it (using OMShell or OMNotebook) for 15 s and plot the variables for the outgoing

Introduction to Modeling and Simulation of Technical and Physical Systems with Modelica,
First Edition. By Peter Fritzson
© 2011 the Institute of Electrical and Electronics Engineers, Inc. Published 2011 by John Wiley & Sons, Inc.

rotational speed on the inertia axis and the voltage on the voltage source (denoted *u* in the figure) in the same plot.

Hint 1: You should look for model components in Modelica. Electrical.Analogue.Basic, Modelica.Electrical.Analogue.Sources, Modelica.Mechanics.Rotational, and so forth.

Hint 2: if you have difficulty finding the names of the variables to plot, you can flatten the model by calling instantiateModel (in OMNotebook or OMShell), which exposes all variable names.

D.2 DC MOTOR WITH SPRING AND INERTIA

Add a torsional spring to the outgoing shaft and another inertia element. Simulate again and look at the results. Adjust some parameters to make a rather stiff spring.

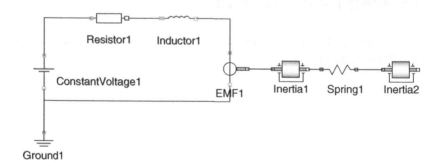

D.3 DC MOTOR WITH CONTROLLER

Add a PI controller to the system and try to control the rotational speed of the outgoing shaft. Verify the result using a step signal for input. Tune the PI controller by changing its parameters in the graphical editor.

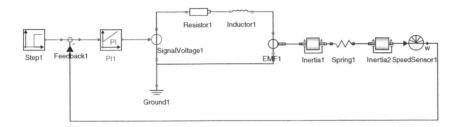

D.4 DC MOTOR AS A GENERATOR

What is needed if you want to make a hybrid DC motor, that is, a DC motor that also can act like a generator for a limited time? Make it work like a DC motor for the first 20 s, then apply a counteracting torque on the outgoing axis for the next 20 s, and then turn off the counteracting torque, that is, you would like to have a torque pulse starting at 20 s and lasting 20 s. Draw the following connection diagram in a graphic model editor, and adjust the starting times and signal height for the Step 1 and Step 2 signal models to get the desired torque pulse.

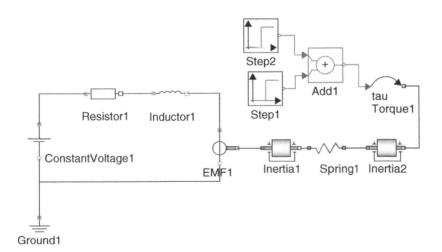

References

Allaby, Michael. Citric acid cycle. *A Dictionary of Plant Sciences*, 1998. http://www.encyclopedia.com/topic/citric_acid.aspx

Allen, Eric, Robert Cartwright and Brian Stoler. DrJava: A Lightweight Pedagogic Environment for Java. In Proceedings of the 33rd ACM Technical Symposium on Computer Science Education (SIGCSE 2002), Cincinnati, Feb. 27–Mar. 3, 2002.

Andersson, Mats. Combined Object-Oriented Modelling in Omola. In Stephenson (ed), Proceedings of the 1992 European Simulation Multiconference (ESM'92), York, UK,Society for Computer Simulation International, June 1992.

Andersson, Mats. Object-Oriented Modeling and Simulation of Hybrid Systems, Ph.D. thesis, Department of Automatic Control, Lund Institute of Technology, Lund, Sweden, 1994.

Arnold, Ken and James Gosling. *The Java Programming Language*, Addison-Wesley, Reading, MA, 1999.

Ashby, W. Ross. *An Introduction to Cybernetics*, Chapman & Hall, London, 1956, p. 39.

Åström, Karl-Johan, Hilding Elmqvist and Sven-Erik Mattsson. Evolution of Continuous-Time Modeling and Simulation. In Zobel and Moeller (eds.), Proceedings of the 12th European Simulation Multiconference (ESM'98), pp. 9–18, Society for Computer Simulation International, Manchester, UK, 1998.

Augustin, Donald C., Mark S. Fineberg, Bruce B. Johnson, Robert N. Linebarger, F. John Sansom and Jon C. Strauss. The SCi Continuous System Simulation Language (CSSL). *Simulation*, 9: 281–303, 1967.

Assmann, Uwe. *Invasive Software Composition*, Springer Verlag, Berlin, 1993.

Bachmann, Bernard (ed.). Proceedings of the 6th International Modelica Conference. Available at www.modelica.org. Bielefeld University, Bielefeld, Germany, March 3–4, 2008.

Birtwistle, G. M., Ole Johan Dahl, B. Myhrhaug and Kristen Nygaard. *SIMULA BEGIN*. Auerbach Publishers, Inc., Boca Raton, FL, 1973.

Brenan, K., S. Campbell and L. Petzold. *Numerical Solution of Initial-Value Problems in Ordinary Differential-Algebraic Equations*. North Holland Publishing, New York, 1989.

Introduction to Modeling and Simulation of Technical and Physical Systems with Modelica,
First Edition. By Peter Fritzson
© 2011 the Institute of Electrical and Electronics Engineers, Inc. Published 2011 by John Wiley & Sons, Inc.

Booch, Grady. *Object Oriented Design with Applications*. Benjamin/Cummings, 1991.

Booch, Grady. *Object-Oriented Analysis and Design*, Addison-Wesley, 1994.

Brück, Dag, Hilding Elmqvist, Sven-Erik Mattsson and Hans Olsson. Dymola for Multi-Engineering Modeling and Simulation. In Proceedings of the 2nd International Modelica Conference, Oberpfaffenhofen, Germany, Mar. 18–19, 2002.

Bunus, Peter, Vadim Engelson and Peter Fritzson. Mechanical Models Translation and Simulation in Modelica. In Proceedings of Modelica Workshop 2000, Lund University, Lund, Sweden, Oct. 23–24, 2000.

Casella, Francesco (ed.). Proceedings of the 7th International Modelica Conference. Available at www.modelica.org. Como, Italy, March 3–4, 2009.

Cellier, Francois E. Combined Continuous/Discrete System Simulation by Use of Digital Computers: Techniques and Tools. Ph.D. thesis, ETH, Zurich, 1979.

Cellier, Francois E., *Continuous System Modelling*, Springer Verlag, Berlin, 1991.

Clauß, Christoph. *Proceedings of the 8th International Modelica Conference*. Available at www.modelica.org. Dresden, Germany, March 20–22, 2011.

Davis, Bill, Horacio Porta and Jerry Uhl. *Calculus & Mathematica Vector Calculus: Measuring in Two and Three Dimensions*. Addison-Wesley, Reading, MA, 1994.

Dynasim AB. Dymola—Dynamic Modeling Laboratory, Users Manual, Version 5.0. Dynasim AB, Lund, Sweden, Changed 2010 to Dassault Systemes, Sweden. www.3ds.com/products/catia/portfolio/dymola, 2003.

Elmqvist, Hilding. A Structured Model Language for Large Continuous Systems. Ph.D. thesis, TFRT-1015, Department of Automatic Control, Lund Institute of Technology, Lund, Sweden, 1978.

Elmqvist, Hilding and Sven-Erik Mattsson. A Graphical Approach to Documentation and Implementation of Control Systems. In Proceedings of the Third IFAC/IFIP Symposium on Software for Computer Control (SOCOCO'82), Madrid, Spain. Pergamon Press, Oxford, 1982.

Elmqvist, Hilding, Francois Cellier and Martin Otter. Object-Oriented Modeling of Hybrid Systems. In Proceedings of the European Simulation Symposium (ESS'93). Society of Computer Simulation, 1993.

Elmqvist, Hilding, Dag Bruck and Martin Otter. Dymola—User's Manual. Dynasim AB, Research Park Ideon, SE-223 70, Lund, Sweden, 1996.

Elmqvist, Hilding, and Sven-Erik Mattsson. Modelica: The Next Generation Modeling Language—An International Design Effort. In Proceedings of First World Congress of System Simulation, Singapore, Sept. 1–3, 1997.

Elmqvist, Hilding, Sven-Erik Mattsson and Martin Otter. Modelica—A Language for Physical System Modeling, Visualization and Interaction. In Proceedings of the 1999 IEEE Symposium on Computer-Aided Control System Design (CACSD'99), Hawaii, Aug. 22–27, 1999.

Elmqvist, Hilding, Martin Otter, Sven-Erik Mattsson and Hans Olsson. Modeling, Simulation, and Optimization with Modelica and Dymola. Book draft, 246 pages. Dynasim AB, Lund, Sweden, Oct. 2002.

Engelson, Vadim, Håkan Larsson and Peter Fritzson. A Design, Simulation, and Visualization Environment for Object-Oriented Mechanical and Multi-Domain Models in Modelica. In Proceedings of the IEEE International Conference on Information Visualization, pp. 188–193, London, July 14–16, 1999.

Ernst, Thilo, Stephan Jähnichen and Matthias Klose. The Architecture of the Smile/M Simulation Environment. In Proceedings 15th IMACS World Congress on Scientific Computation, Modelling and Applied Mathematics, Vol. 6, Berlin, Germany, pp. 653–658. See also http://www.first.gmd.de/smile/smile0.html, 1997.

Fauvel, John, Raymond Flood, Michael Shortland and Robin Wilson. *LET NEWTON BE! A New Perspective on His Life and Works, Second Edition*. Oxford University Press, Oxford, 1990.

Felleisen, Matthias, Robert Bruce Findler, Matthew Flatt and Shiram Krishnamurthi. The DrScheme Project: An Overview. In Proceedings of the ACM SIGPLAN 1998 Conference on Programming Language Design and Implementation (PLDI'98), Montreal, Canada, June 17–19, 1998.

Fritzson, Dag and Patrik Nordling. Solving Ordinary Differential Equations on Parallel Computers Applied to Dynamic Rolling Bearing Simulation. In *Parallel Programming and Applications*, P. Fritzson, and L. Finmo (eds.), IOS Press, 1995.

Fritzson, Peter and Karl-Fredrik Berggren. Pseudo-Potential Calculations for Expanded Crystalline Mercury, *Journal of Solid State Physics*, 1976.

Fritzson, Peter. Towards a Distributed Programming Environment based on Incremental Compilation. Ph.D. thesis, 161 pages. Dissertation no. 109, Linköping University, Apr. 13, 1984.

Fritzson, Peter and Dag Fritzson. The Need for High-Level Programming Support in Scientific Computing—Applied to Mechanical Analysis. *Computers and Structures*, 45(2): 387–295, 1992.

Fritzson, Peter, Lars Viklund, Johan Herber and Dag Fritzson. Industrial Application of Object-Oriented Mathematical Modeling and Computer Algebra in Mechanical Analysis, In Proceedings of TOOLS EUROPE'92, Dortmund, Germany, Mar. 30–Apr. 2. Prentice Hall, 1992.

Fritzson, Peter, Lars Viklund, Dag Fritzson and Johan Herber. High Level Mathematical Modeling and Programming in Scientific Computing, *IEEE Software*, pp. 77–87, July 1995.

Fritzson, Peter and Vadim Engelson. Modelica—A Unified Object-Oriented Language for System Modeling and Simulation. Proceedings of the 12th European Conference on Object-Oriented Programming (ECOOP'98), Brussels, Belgium, July 20–24, 1998.

Fritzson, Peter, Vadim Engelson and Johan Gunnarsson. An Integrated Modelica Environment for Modeling, Documentation and Simulation. In Proceedings of Summer Computer Simulation Conference '98, Reno, Nevada, July 19–22, 1998.

Fritzson, Peter (ed.). Proceedings of SIMS'99—The 1999 Conference of the Scandinavian Simulation Society, Linköping, Sweden, Oct. 18–19, 1999. Available at www.scansims.org.

Fritzson, Peter (ed.). Proceedings of Modelica 2000 Workshop, Lund University, Lund, Sweden, Oct. 23–24, 2000. Available at www.modelica.org.

Fritzson, Peter and Peter Bunus. Modelica—A General Object-Oriented Language for Continuous and Discrete-Event System Modeling and Simulation. Proceedings of the 35th Annual Simulation Symposium, San Diego, California, Apr. 14–18. 2002.

Fritzson, Peter, Peter Aronsson, Peter Bunus, Vadim Engelson, Henrik Johansson, Andreas Karström and Levon Saldamli. The Open Source Modelica Project. In Proceedings of the 2nd International Modelica Conference, Oberpfaffenhofen, Germany, Mar. 18–19, 2002.

Fritzson Peter, Mats Jirstrand and Johan Gunnarsson. MathModelica—An Extensible Modeling and Simulation Environment with Integrated Graphics and Literate Programming. In Proceedings of the 2nd International Modelica Conference, Oberpfaffenhofen, Germany, Mar. 18–19, 2002. Available at www.ida.liu.se/labs/pelab/modelica/ and at www.modelica.org.

Fritzson, Peter (ed.). Proceedings of the 3rd International Modelica Conference. Linköping University, Linköping, Sweden, Nov 3–4, 2003. Available at www.modelica.org.

Fritzson Peter. *Principles of Object Oriented Modeling and Simulation with Modelica 2.1*, Wiley-IEEE Press, Hoboken, NJ, 2004

Fritzson Peter, Peter Aronsson, Håkan Lundvall, Kaj Nyström, Adrian Pop, Levon Saldamli and David Broman. The OpenModelica Modeling, Simulation, and Software Development Environment. In *Simulation News Europe*, 44/45, December 2005. See also: http://www.openmodelica.org

Fritzson, Peter. MathModelica—An Object Oriented Mathematical Modeling and Simulation Environment. *Mathematica Journal* 10(1), February. 2006.

Fritzson, Peter. Electronic Supplementary Material to Introduction to Modeling and Simulation of Technical and Physical Systems with Modelica. www.openmodelica.org, July 2011.

Gottwald, S., W. Gellert (Contributor), and H. Kuestner (Contributor). *The VNR Concise Encyclopedia of Mathematics*. Second edition, Van Nostrand Reinhold, New York, 1989.

Hairer, E., S. P. Nørsett and G. Wanner. *Solving Ordinary Differential Equations I. Nonstiff Problems*, Second Edition. Springer Series in Computational Mathematics, Springer Verlag, Berlin, 1992.

Hairer, E. and G. Wanner. *Solving Ordinary Differential Equations II. Stiff and Differential-Algebraic Problems*. Springer Series in Computational Mathematics, Springer Verlag, Berlin, 1991.

Hyötyniemi, Heikki. Towards New Languages for Systems Modeling. In Proceedings of SIMS 2002, Oulu, Finland, Sept. 26–27, 2002, Available at www.scansims.org.

IEEE. Std 610.3–1989. *IEEE Standard Glossary of Modeling and Simulation Terminology*, 1989.

IEEE Std 610.12–1990. *IEEE Standard Glossary of Software Engineering Terminology*, 1990.

IEEE Std 1076.1–1999. IEEE Computer Society Design Automation Standards Committee, USA. *IEEE Standard VHDL Analog and Mixed-Signal Extensions*, Dec. 23, 1999.

Knuth, Donald E. Literate Programming. *The Computer Journal*, 27(2): 97–111, May 1984.

Kral, Christian and Anton Haumer. Proceedings of the 6th International Modelica Conference. Available at www.modelica.org. Vienna, Austria, Sept 4–6, 2006.

Kågedal, David and Peter Fritzson. Generating a Modelica Compiler from Natural Semantics Specifications. In Proceedings of the Summer Computer Simulation Conference'98, Reno, Nevada, July 19–22 1998.

Lengquist Sandelin, Eva-Lena and Susanna Monemar. DrModelica—An Experimental Computer-Based Teaching Material for Modelica. Master's thesis, LITH-IDA-Ex03/3, Department of Computer and Information Science, Linköping University, Linköping, Sweden, 2003.

Lengquist Sandelin, Eva-Lena, Susanna Monemar, Peter Fritzson and Peter Bunus. DrModelica— A Web-Based Teaching Environment for Modelica. In Proceedings of the 44th Scandinavian Conference on Simulation and Modeling (SIMS'2003),Västerås, Sweden, Sept. 18–19, 2003. Available at www.scansims.org.

Ljung, Lennart and Torkel Glad. *Modeling of Dynamic Systems*, Prentice Hall, 1994.

MathCore Engineering AB. Home page: www.mathcore.com. MathCore Engineering AB, Linköping, Sweden, 2003.

MathWorks Inc. *Simulink User's Guide*, 2001

MathWorks Inc. *MATLAB User's Guide*, 2002.

Mattsson, Sven-Erik, Mats Andersson and Karl-Johan Åström. Object-Oriented Modelling and Simulation. In Linkens (ed.), *CAD for Control Systems*, Chapter 2, pp. 31–69. Marcel Dekker, New York, 1993.

Meyer, Bertrand. *Object-Oriented Software Construction*, Second Edition, Prentice-Hall, Englewood Cliffs, 1997.

Mitchell, Edward E. L. and Joseph S. Gauthier. *ACSL: Advanced Continuous Simulation Language—User Guide and Reference Manual*. Mitchell & Gauthier Assoc., Concord, Mass, 1986.

Modelica Assocation. Home page: www.modelica.org. Last accessed 2010.

Modelica Association. Modelica—A Unified Object-Oriented Language for Physical Systems Modeling: Tutorial and Design Rationale Version 1.0, Sept. 1997.

Modelica Association. Modelica—A Unified Object-Oriented Language for Physical Systems Modeling: Tutorial, Version 1.4., Dec. 15, 2000. Available at http://www.modelica.org

Modelica Association. Modelica—A Unified Object-Oriented Language for Physical Systems Modeling: Language Specification Version 3.2., March 2010. Available at http://www.modelica.org

ObjectMath Home page: http://www.ida.liu.se/labs/pelab/omath.

OpenModelica page: http://www.openmodelica.org.

Otter, Martin, Hilding Elmqvist, and Francois Cellier. Modeling of Multibody Systems with the Object-Oriented Modeling Language Dymola, Nonlinear Dynamics, 9, pp. 91–112. Kluwer Academic Publishers, 1996.

Otter, Martin. Objektorientierte Modellierung Physikalischer Systeme, Teil 1: Übersicht. In *Automatisierungstechnik*, 47(1): A1–A4. 1999. In German, the first in a series of 17 articles, 1999.

Otter, Martin (ed.) Proceedings of the 2nd International Modelica Conference. Available at www.modelica.org. Oberpfaffenhofen, Germany, Mar. 18–19, 2002.

Otter, Martin, Hilding Elmqvist and Sven-Erik Mattsson. The New Modelica Multi-Body Library. Proceedings of the 3rd International Modelica ConferenceLinköping, Sweden, Nov 3–4, 2003. Available at www.modelica.org

PELAB. Page on Modelica Research at PELAB, Programming Environment Laboratory, Dept. of Computer and Information Science, Linköping University, Sweden, 2003. Available at www.ida.liu.se/labs/pelab/modelica,

Piela, P. C., T. G. Epperly, K. M. Westerberg, and A. W. Westerberg. ASCEND—An Object-Oriented Computer Environment for Modeling and Analysis: The Modeling Language, *Computers and Chemical Engineering*, 15(1): 53–72, 1991. Web page: (http://www.cs.cmu.edu/~ascend/Home.html)

Pritsker, A. and B. Alan. *The GASP IV Simulation Language*. Wiley, New York, 1974.

Rumbaugh, J. M., M. Blaha, W. Premerlain, F. Eddy and W. Lorensen. *Object Oriented Modeling and Design*. Prentice-Hall, 1991.

Sahlin, Per. and E. F. Sowell. A Neutral Format for Building Simulation Models. In Proceedings of the Conference on Building Simulation, IBPSA, Vancouver, Canada, 1989.

Sargent, R. W. H. and Westerberg, A. W. *Speed-Up in Chemical Engineering Design*, Transaction Institute in Chemical Engineering, 1964.

Schmitz, Gerhard (ed.). Proceedings of the 4th International Modelica Conference. Available at www.modelica.org. Technical University Hamburg-Harburg, Germany, March 7–8, 2005.

Shumate, Ken and Marilyn Keller. *Software Specification and Design: A Disciplined Approach for Real-Time Systems*. Wiley, New York, 1992.

Stevens, Richard, Peter Brook, Ken Jackson and Stuart Arnold. Systems *Engineering: Coping with Complexity*. Prentice-Hall, London, 1998.

Szyperski, Clemens. *Component Software—Beyond Object-Oriented Programming*. Addison-Wesley, Reading, MA, 1997.

Tiller, Michael. *Introduction to Physical Modeling with Modelica*. Kluwer, Amsterdam, 2001.

Viklund Lars, and Peter Fritzson. ObjectMath—An Object-Oriented Language and Environment for Symbolic and Numerical Processing in Scientific Computing, *Scientific Programming*, 4: 229–250, 1995.

Wolfram, Stephen. *The Mathematica Book*. Wolfram Media Inc, 1997.

Wolfram, Stephen. *A New Kind of Science*. Wolfram Media Inc, 2002.

Index

Introduction to Modeling and Simulation of Technical and Physical Systems with Modelica,
First Edition. By Peter Fritzson
© 2011 the Institute of Electrical and Electronics Engineers, Inc. Published 2011 by John Wiley & Sons, Inc.